A Biologist's Guide to
ANALYSIS OF DNA
MICROARRAY DATA

A Biologist's Guide to *ANALYSIS OF DNA MICROARRAY DATA*

Steen Knudsen
Center for Biological Sequence Analysis
BioCentrum-DTU

Technical University of Denmark

A JOHN WILEY & SONS, INC., PUBLICATION

Cover art supplied courtesy of www.molmine.com

This text is printed on acid-free paper. ∞

For ordering and customer service, call 1-800-CALL-WILEY.

Library of Congress Cataloging-in-Publication Data:

Library of Congress Cataloging-in-Publication Data is available.
ISBN 0-471-22490-1

Printed in the United States of America

10 9 8 7 6 5 4 3 2 1

To Tarja

Contents

Preface

I am often asked, "Do you have a good text I can read on analysis of DNA array data?" This is an attempt at providing such a text for students and scientists alike who venture into the field of DNA array data analysis for the first time. The book is written for biologists without special training in data analysis and statistics. Mathematical stringency is sacrificed for intuitive and visual introduction of concepts. Simple solutions are emphasized where more complicated solutions may be more correct. Methods are introduced by simple examples and citations of relevant literature. Practical computer solutions to common analysis problems are suggested, with an emphasis on software developed at and made freely available by my own lab. The text emphasizes gene expression analysis.

This text takes over where the image analysis software that usually comes with DNA array equipment leaves you: with a file of signal intensities and fold change compared to a control. The information in that file will prompt questions such as: How is it scaled? What is the error in the data? When can I say that a certain gene is up-regulated? What do I do with the thousands of genes that show some regulation? How much information can I get out of my data? This text will attempt to answer those questions and others that will come into mind as you delve further into the data.

Each chapter has a section on Further Reading that categorizes relevant literature by topic. In addition, all literature is categorized by author in the Bibliography section.

STEEN KNUDSEN

Lyngby, Denmark, December 2001

Acknowledgments

Christopher Workman and Laurent Gautier inspired me for many aspects of this book and also implemented many methods used in the book.

I thank Nikolaj Blom, Christopher Workman, and Henrik Bjørn Nielsen for helpful suggestions on the manuscript.

I thank my collaborators Claus Nielsen, Kenneth Thirstrup, Torben Ørntoft, Friedrik Wikman, Thomas Thykjaer, Mogens Kruhøffer, Karin Demtröder, Hans Wolf, Lars Dyrskjøt Andersen, Casper Møller Frederiksen, Jeppe Spicker, Lars Juhl Jensen, Carsten Friis, Hanne Jarmer, Hans-Henrik Saxild, Randy Berka, Matthew Piper, Steen Westergaard, Christoffer Bro, Thomas Jensen, and Kristine Dahlin for allowing me to use examples generated from our collaborative research.

I am grateful to Center Director Søren Brunak for creating the environment, and to the The Danish National Research Foundation, the Danish Biotechnology Instrument Center and Novozymes A/S for funding the research, that made this book possible.

S. K.

1

Introduction

1.1 HYBRIDIZATION

The fundamental basis of DNA microarrays is the process of *hybridization*. Two DNA strands hybridize if they are complementary to each other. One or both strands of the DNA hybrid can be replaced by RNA and hybridization will still occur as long as there is complementarity.

Hybridization has for decades been used in molecular biology as the basis for such techniques as Southern blotting and Northern blotting. In Southern blotting, a small string of DNA, an *oligonucleotide*, is used to hybridize to complementary fragments of DNA that have been separated according to size in a gel electrophoresis. If the oligonucleotide is radioactively labeled, the hybridization can be visualized on a photographic film that is sensitive to radiation. In Northern blotting, a radiolabeled oligonucleotide is used to hybridize to messenger RNA that has been run through a gel. If the oligo is specific to a single messenger RNA, then it will bind to the location (*band*) of that messenger in the gel. The amount of radiation captured on a photographic film is dependent to some extent on the amount of radiolabeled probe present in the band, which again is dependent on the amount of messenger. So this method is a semiquantitative detection of individual messengers.

DNA arrays are a massively parallel version of Northern and Southern blotting. Instead of distributing the oligonucleotide probes over a gel containing samples of RNA or DNA, the oligonucleotide probes have been immobilized on a surface. They can be immobilized at micrometer distances, so it is possible to place many different oligonucleotide probes on a small single surface of

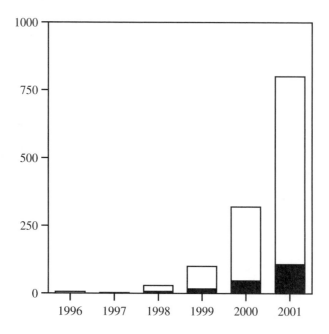

Fig. 1.1 The number of published papers referring to DNA microarrays. Open bars : Medline search for the the word "microarray." Closed bars : Number of papers citing Affymetrix GeneChip technology.

one square centimeter. Instead of radioactive labeling, the sample is usually labeled with a fluorescent dye that can be detected by a light scanner that scans the surface of the chip for hybridized material.

Where before it was possible to run a couple of Northern blots or a couple of Southern blots in a day, it is now possible with DNA arrays to run tens of thousands of hybridizations. This has in some sense revolutionized molecular biology and medicine. Instead of studying one gene and one messenger at a time, experimentalists are now studying many genes and many messengers at the same time. In fact, DNA arrays are often used to study *all* known messengers of an organism. This has opened the possibility of an entirely new, systemic view of how cells react in response to certain stimuli. It is also an entirely new way to study human disease by viewing how it affects the expression of all genes inside the cell. Figure 1.1 illustrates the revolution of DNA arrays in biology and medicine by the number of papers published on the topic.

When DNA microarrays are used for measuring the concentration of messenger RNA in living cells, a *probe* of one DNA strand that matches a particular messenger RNA in the cell is used. The concentration of a particular messenger is a result of *expression* of its corresponding gene, so this application is often referred to as *expression analysis*. When different probes matching all messenger RNAs in a cell are used, a snapshot of the total messenger RNA pool of a living cell or tissue can be obtained. This is often referred to as an *expression profile* because it reflects the expression of every single measured gene at that particular moment. Expression profile is also sometimes used to describe the expression of a single gene over a number of conditions.

Expression analysis can also be performed by a method called *serial analysis of gene expression (SAGE)*. Instead of using microarrays, SAGE relies on traditional DNA sequencing to identify and enumerate the number of messenger RNAs in a cell (Section 1.5).

Another traditional application of DNA microarrays is to detect mutation in specific genes. The massively parallel nature of DNA microarrays allows the simultaneous screening of many, if not all, possible mutations within a single gene. This is referred to as *genotyping* (Section 11).

The treatment of array data does not depend so much on the technology used to gather the data as it depends on the application in question. Genotyping and expression analysis are two completely different applications, and they will be treated separately in this text. Most of the information will address analysis of expression data, and a separate chapter will address genotyping chips.

For expression analysis there are two major technologies available. There is the Affymetrix, Inc. GeneChip system, which uses prefabricated oligonucleotide chips (Figure 1.2 and 1.3), and there are custom made chips where a robot is used to spot cDNA, oligonucleotides, or PCR products on a glass slide (Figure 1.4).

1.2 AFFYMETRIX GENECHIP TECHNOLOGY

Affymetrix uses equipment similar to that which is used for making silicon chips for computers, and thus allows mass production of very large chips at reasonable cost. Where computer chips are made by creating masks that control a photolithographic process for removal or deposit of material on the chip surface, Affymetrix uses masks to control synthesis of oligonucleotides on the surface of a chip. The standard phosphoramidite method for synthesis of oligonucleotides has been modified to allow light control of the individual steps. The masks control the synthesis of several hundred thousand squares, each containing many copies of an oligo. So the result is several hundred thousand different oligos, each of them present in millions of copies.

That large number of oligos, up to 25 nucleotides long, has turned out to be very useful as an experimental tool to replace all experimental detection procedures that in the past relied on using oligonuclotides: Southern, Northern, and dot blotting as well as sequence specific probing and mutation detection.

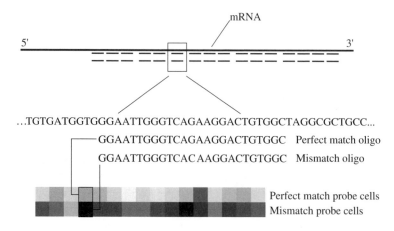

Fig. 1.2 The Affymetrix GeneChip technology. The presence of messenger RNA is detected by a series of probe pairs that differ in only one nucleotide. Hybridization of fluorescent messenger RNA to these probe pairs on the chip is detected by laser scanning of the chip surface. (Figure by Christoffer Bro.)

For expression analysis, up to 40 oligos are used for the detection of each gene. Affymetrix has chosen a region of each gene that (presumably) has the least similarity to other genes. From this region 11–20 oligos are chosen as perfect matches (PM) (i.e., perfectly complementary to the mRNA of that gene). In addition, they have generated 11–20 mismatch oligos (MM), which are identical to the PM oligos except for the central position 13, where one nucleotide has been changed to its complementary nucleotide. Affymetrix claims that the MM oligos will be able to detect nonspecific and background hybridization, which is important for quantifying weakly expressed mRNAs. However, for weakly expressed mRNAs where the signal-to-noise ratio is smallest, subtracting mismatch from perfect match adds considerably to the noise in the data (Schadt, et al., 2000). That is because subtracting one noisy number from another noisy number yields a third number with even more noise.

The hybridization of each oligo to its target depends on its sequence. All 11–20 PM oligos for each gene have different sequence, so the hybrization will not be uniform. That is of no consequence as long as we wish to detect only *changes* in mRNA concentration between experiments. How such a change is calculated from the intensities of the 22–40 probes for each genes will be covered in Section 3.1.

In order to detect hybridization of a target mRNA by a probe on the chip, the target mRNA needs to be labeled with a fluorochrome. As shown in Figure 1.3, the steps from cell to chip usually are as follows:

- Extract total RNA from cell (usually using TRIzol or RNeasy kits).

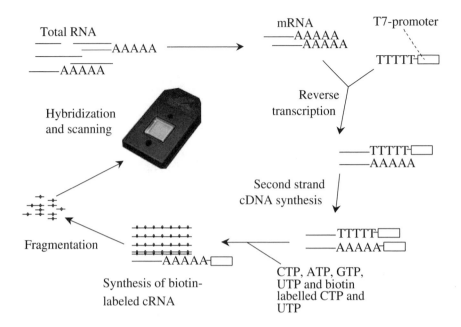

Fig. 1.3 Preparation of sample for GeneChip arrays. Messenger RNA is extracted from the cell and converted to cDNA. It then undergoes an amplification and labeling step before fragmentation and hybridization to 25-mer oligos on the surface of the chip. After washing of unhybridized material, the chip is scanned in a confocal laser scanner and the image analyzed by computer. (Figure by Christoffer Bro.)

- Convert mRNA to cDNA using reverse transcriptase and a poly-T primer.

- Amplify resulting cDNA using T7 RNA polymerase in the presence of biotin-UTP and biotin-CTP, so each cDNA will yield 50–100 copies of biotin-labeled cRNA.

- Incubate cRNA at 94 degrees in fragmentation buffer to produce cRNA fragments of length 35 to 200 nucleotides.

- Hybridize to chip and wash away non-hybridized material.

- Stain hybridized biotin-labeled cRNA with Streptavidin-Phycoerythrin and wash.

- Scan chip in confocal laser scanner.

- Amplify the signal on the chip with goat IgG and biotinylated antibody.

- Scan chip in scanner again.

Table 1.1 Performance of the Affymetrix GeneChip technology. Numbers refer to chips in routine use and the current limit of the technology (Lipshutz, et al., 1999; Baugh, et al., 2001).

	Routine use	Current limit
Starting material	5 μg total RNA	2 ng total RNA
Detection specificity	$1 : 10^5$	$1 : 10^6$
Difference detection	twofold changes	10% changes
Discrimination of related genes	70-80% identity	93% identity
Dynamic range (linear detection)	3 orders of magn.	4 orders of magn.
Probe pairs per gene	20	4
Number of genes per array	12,000	40,000

Usually, 5 to 10 μg of total RNA are required for the procedure. But new improvements to the cDNA synthesis protocols reduce the required amount to 100 ng. If two cycles of cDNA synthesis and cRNA synthesis are performed, the detection limit can be reduced to 2 ng of total RNA (Baugh, et al., 2001). The current performance of the Affymetrix GeneChip technology is summarized in Table 1.1.

1.3 SPOTTED ARRAYS

In another major technology, spotted arrays, a robot spotter is used to move small quantities of probe in solution from a microtiter plate to the surface of a glass plate. The probe can consist of cDNA, PCR product, or oligonucleotides. Each probe is complementary to a unique gene. Probes can be fixed to the surface in a number of ways. The classical way is by non-specific binding to polylysine-coated slides. The steps involved in making the slides can be summarized as follows (Figure 1.4):

- Coat glass slides with polylysine.

- Prepare probes in microtiter plates.

- Use robot to spot probes on glass slides.

- Block remaining exposed amines of polylysine with succinic anhydride.

- Denature DNA (if double-stranded) by heat.

The steps involved in preparation of sample and hybridizing to the array can be summarized as follows (Figure 1.4):

- Extract total RNA from cells.

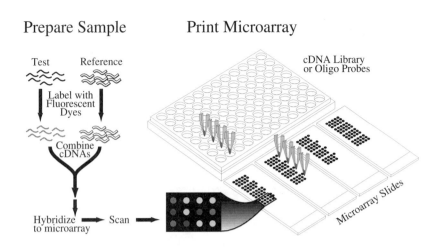

Fig. 1.4   The spotted array technology. A robot is used to transfer probes in solution from a microtiter plate to a glass slide where they are dried. Extracted mRNA from cells is converted to cDNA and labeled fluorescently. Sample is labeled red and control is labeled green. After mixing, they are hybridized to the probes on the glass slide. After washing away unhybridized material, the chip is scanned with a confocal laser and the image analyzed by computer. (See color plate.)

- Optional: isolate mRNA by polyA tail.

- Convert to cDNA in the presence of Aminoallyl-dUTP (AA-dUTP).

- Label with Cy3 or Cy5 fluorescent dye linking to AA-dUTP.

- Hybridize labeled mRNA to glass slides.

- Wash unhybridized material away.

- Scan slide and analyze image.

The advantage compared to Affymetrix GeneChips is that you can design any probe for spotting on the array. The disadvantage is that spotting will not be nearly as uniform as the *in situ* synthesized Affymetrix chips and that the cost of oligos, for chips containing thousands of probes, becomes high. From a data analysis point of view the main difference is that in the cDNA array usually the sample and the control are hybridized to the same chip using different fluorochromes, whereas the Affymetrix chip can handle only one fluorochrome so two chips are required to compare a sample and a control. Table 1.2 shows the current performance of the spotted array technology

Table 1.2   Performance of the spotted array technology (Schena, 2000).

	Routine use
Starting material	10-20 μg total RNA
Dynamic range (linear detection)	3 orders of magnitude
Number of probes per gene	1
Number of genes or ESTs per array	$\approx$10,000

1.4   SERIAL ANALYSIS OF GENE EXPRESSION (SAGE)

A technology that is both widespread and attractive because it can be run on a standard DNA sequencing apparatus is serial analysis of gene expression (SAGE) (Velculescu, et al., 1995; Yamamoto, et al., 2001). In SAGE, cDNA fragments, "tags," are concatenated by ligation and sequenced. The number of times a tag occurs, and is sequenced, is related to the abundance of its corresponding messenger. Thus, if enough concatenated tags are sequenced one can get a quantitative measure of the mRNA pool. Bioinformatics first enters the picture when one wishes to find the gene corresponding to a particular tag, which may be only 9–14 bp long. Each tag is searched against a database (van Kampen, et al., 2000; Margulies, et al., 2000; Lash, et al., 2000) to find one or more genes that match.

The steps involved in the SAGE methods can be summarized as follows (see also Figure 1.5):

- Extract RNA and convert to cDNA using biotinylated poly-T primer.

- Cleave with a frequently cutting (4 bp recognition site) restriction enzyme (anchoring enzyme).

- Isolate 3'-most restriction fragment with biotin-binding streptavidin-coated beads.

- Ligate to linker which contains a type IIS restriction site for tagging enzyme. (cuts 20 bp away from site) and primer sequence.

- Cleave with tagging enzyme that cuts up to 20 bp away from recognition site.

- Ligate and amplify with primers complementary to linker.

- Cleave with anchoring enzyme, isolate ditags.

- Concatenate and clone.

- Sequence clones.

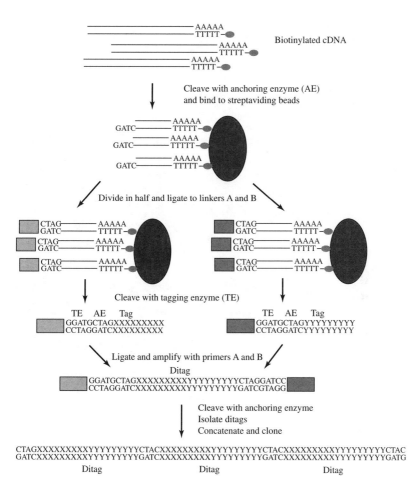

Fig. 1.5 Schematic overview of SAGE methods (based on Velculescu, et al. 1995). (See color plate.)

The analysis of SAGE data is similar to the analysis of array data described through out this book except that the statistical analysis of significance is different (Man, et al., 2000; Lash, et al., 2000; Audic, et al., 1997).

1.5 EXAMPLE: AFFYMETRIX VS. SPOTTED ARRAYS

Our lab has in a collaboration (Knudsen, et al., 2001) performed both cDNA array analysis and Affymetrix chip analysis of human T cells infected with Human Immunodeficiency Virus (HIV). Figure 1.6 shows mRNA extracted from the T cells and visualized with a cDNA array. First, the mRNA was

Fig. 1.6 cDNA microarray of genes affected by HIV infection. (See color plate.)

converted to cDNA and then it was labeled with a fluorochrome. We used a red fluorochrome for the mRNA that was extracted from the HIV-infected cells ("experiment," "sample," or "treatment") and we used a green fluorochrome for mRNA that was extracted from the noninfected cells ("control"). Because we used different fluorochromes we could apply both sample and control to the same chip where we have already spotted probes for the genes we were interested in. Figure 1.6 shows such genes.

After hybridization and washing, the chip was scanned and the image processed by a computer. We can now deduce the ratio between the expression of each gene in HIV-infected cells and the gene in uninfected cells as the ratio between the intensity of red and green color. If the color is yellow, there is no change. If it is red, there is an upregulation; if it is green there is a downregulation (Figure 1.6). We can also estimate mRNA concentration from the intensity of the spot.

We took the same mRNA, processed it, and put it on an Affymetrix chip with 6800 human genes. Figure 1.7 shows part of the surface of this chip that probes for just one gene. Before putting it on the chip, the mRNA was

Fig. 1.7 Part of Affymetrix chip probing one gene affected by HIV infection.

converted to cDNA and an *in vitro* transcription step was used to amplify the amount of mRNA. After fragmentation and labeling with a fluorescent dye, the sample was hybridized to the chip surface where a total of 40 oligonucleotide probes of length 25 are used to detect the presence and concentration of each gene messenger. Twenty oligos are chosen from different areas of the gene. Each of these oligos is called a perfect match (PM). For each PM oligo there is an identical oligo with one mismatch at the center position of the 25-mer. This mismatch (MM) oligo is included to compensate for nonspecific hybridization as well as cross-hybridization.

In Figure 1.7 the PM oligos and MM oligos are shown on top of each other. Note that Affymetrix does not use a two-color system, so we have to run one chip for the sample and one chip for the control. The conditions for comparing those two chips will be described in a later section.

1.6 SUMMARY

Spotted arrays are made by deposition of a probe on a solid support. Affymetrix chips are made by light mask technology. The latter is easier to control and therefore the variation between chips is smaller in the latter technology. Spotted arrays offer more flexibility, however. Data analysis does not differ much between the two types of arrays.

Serial analysis of gene expression (SAGE) is yet another method for analyzing the abundance of mRNA by sequencing concatenated fragments of their corresponding cDNA. The number of times a cDNA fragment occurs in the concatenated sequence is proportional to the abundance of its corresponding messenger.

Figure 1.8 shows a schematic overview of how one starts with cells in two different conditions (with and without HIV virus, for example) and ends up with mRNA from each condition hybridized to a DNA array. Also shown is a traditional approach for spotting differences in mRNA populations, *differential display*. This method, which relies on gel electrophoresis of radiolabled fragments, gives only a qualitative assessment of differences in mRNA expression and for that reason it will not be covered further in this book.

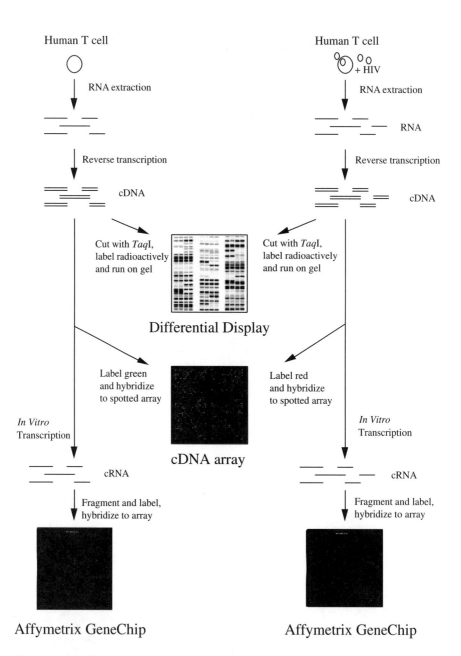

Differential Display

cDNA array

Affymetrix GeneChip   Affymetrix GeneChip

Fig. 1.8 Overview of methods for comparing mRNA populations in cells from two different conditions. (See color plate.)

1.7 FURTHER READING

Knudsen, S., Nielsen, H.B., Nielsen, C., Thirstrup, K., Blom, N., Sicheritz-Ponten, T., Gautier, L., Workman, C., and Brunak, S. (2001). T cell transcriptional responses to HIV infection in vitro. Manuscript in preparation.

Overview as well as details of Affymetrix technology

Affymetrix (1999). *GeneChip Analysis Suite.* User Guide, version 3.3.

Affymetrix (2000). *GeneChip Expression Analysis.* Technical Manual.

Baugh, L. R., Hill, A. A., Brown, E. L., and Hunter, C. P. (2001). Quantitative analysis of mRNA amplification by in vitro transcription. *Nucleic Acids Research* 29:E29.

Lipshutz, R. J., Fodor, S. P. A., Gingeras, T. R., and Lockhart, D. J. (1999). High density synthetic oligonucleotide arrays. *Nature Genetics Chipping Forecast* 21:20–24.

Lockhart, D. J., Dong, H., Byrne, M. C., Follettie, M. T., Gallo, M. V., Chee, M. S., Mittmann, M., Wang C., Kobayashi, M., Horton, H., and Brown, E. L. (1996). Expression monitoring by hybridization to high-density oligonucleotide arrays. *Nature Biotechnology* 14:1675–1680.

Wodicka, L., Dong, H., Mittmann, M., Ho, M. H., and Lockhart, D. J. (1997). Genome-wide expression monitoring in Saccharomyces cerevisiae. *Nature Biotechnology* 15:1359–1367.

Overview as well as details of spotted arrays

Pat Brown lab website.[1]

Schena, Mark. (2000). *Microarray Biochip Technology.* Sunnyvale, CA: Eaton.

Schena, Mark. (1999). *DNA microarrays: A practical approach* (Practical Approach Series, 205). Oxford: Oxford Univ. Press.

Microarrays web site.[2] Includes protocols largely derived from the Cold Spring Harbor Laboratory Microarray Course manual.

---

[1]http://brownlab.stanford.edu
[2]http://www.microarrays.org/

Differential display

Matz, M., Usman, N., Shagin, D., Bogdanova, E., and Lukyanov, S. (1997). Ordered differential display: A simple method for systematic comparison of gene expression profiles. *Nucleic Acids Res.* 25:2541–2.

Serial analysis of gene expression

Audic, S., and Claverie, J. M. (1997). The significance of digital gene expression profiles. *Genome Res.* 7:986–995.

van Kampen, A. H., van Schaik, B. D., Pauws, E., Michiels, E. M., Ruijter, J. M., Caron, H. N., Versteeg, R., Heisterkamp, S. H., Leunissen, J. A., Baas, F., and van der Mee, M. (2000). USAGE: A web-based approach towards the analysis of SAGE data. *Bioinformatics.* 16:899–905.

Lash, A. E., Tolstoshev, C. M., Wagner, L., Schuler, G. D., Strausberg, R. L., Riggins, G. J., and Altschul, S. F. (2000). SAGEmap: A public gene expression resource. *Genome Res.* 10:1051–60.

Man, M. Z., Wang, X., and Wang, Y. (2000). POWER_SAGE: Comparing statistical tests for SAGE experiments. *Bioinformatics.* 16:953–9.

Margulies, E. H., and Innis, J. W. (2000). eSAGE: Managing and analysing data generated with serial analysis of gene expression (SAGE). *Bioinformatics.* 16:650–1.

Velculescu, V. E., Zhang, L., Vogelstein, B., and Kinzler, K. W. (1995). Serial analysis of gene expression. *Science.* 270:484–7.

Yamamoto, M., Wakatsuki, T., Hada, A., and Ryo, A. (2001). Use of serial analysis of gene expression (SAGE) technology. *J Immunol. Methods.* 250:45–66. Review.

2

Overview of Data Analysis

Figure 2.1 represents an overview of the general data analysis methods presented in this book. In particular it shows the order in which to apply them and which methods to choose in different situations. For clarity, not all possible orders of analysis have been shown. For example, PCA and clustering can be performed after an ANOVA or *t*-test.

See Table 2.1 for where to find details of the individual methods.

Table 2.1 Section and page number for methods shown in Figure 2.1.

Method	Section	Page
Scaling	3.2	18
t-test	3.5	22
ANOVA	3.5	22
PCA	4.1	33
Clustering	5	41
Fold change	3.4	21
Promoter analysis	6.2	56
Function prediction	6.1	55
Classification	8	75

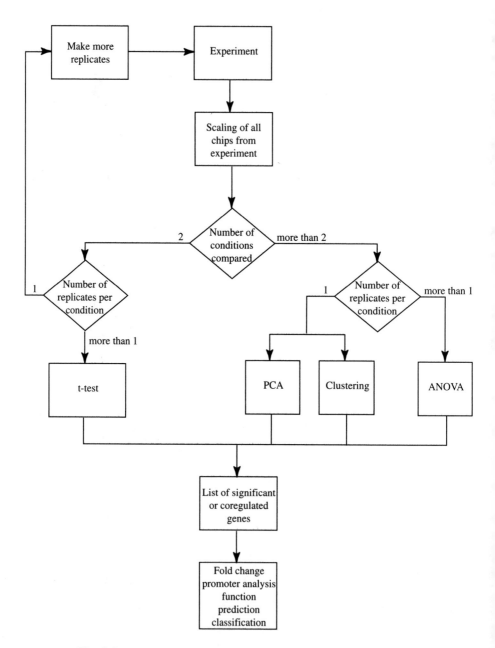

Fig. 2.1   Overview of data analysis methods presented in this book.

3

Basic Data Analysis

In gene expression analysis, technological problems and biological variation make it difficult to distinguish signal from noise. Once we obtain reliable data, we look for patterns and need to determine their significance.

—Vingron, 2001

3.1 ABSOLUTE MEASUREMENTS

The image processing software that comes with your DNA array equipment identifies the probe cells, calculates their signal intensities, subtracts background, and measures the noise level. For the cDNA array you are then given the intensity level for each probe cell (spot) in integer numbers for the red and green channel respectively.

Affymetrix, in the early versions of their software, calculated an Average Difference between probe pairs. The Affymetrix chip has several probe pairs for each gene. A probe pair consists of a perfect match (PM) oligo and a mismatch (MM) oligo for comparison. The mismatch oligo differs from the perfect match oligo in only one position and is used to detect nonspecific hybridization. *AvgDiff* was calculated as follows:

$$AvgDiff = \frac{\sum_N PM - MM}{N}$$

where N is the number of probe pairs used for the calculation (probe pairs which deviate by more than 3 standard deviations from the mean are *excluded* from the calculation). If the AvgDiff number is negative or very small, it means that either the target is absent or there is nonspecific hybridization. Affymetrix calculates an Absolute Call based on probe statistics: Absent, Marginal, or Present (refer to Affymetrix manual for the decision matrix used for making the Absolute Call). In new versions of their software, Affymetrix has replaced AvgDiff with a Signal which is calculated as described in a statistical algorithms reference guide.[1]

Li and Wong (2001a, b) instead calculate a weighted average difference:

$$\tilde{\theta} = \frac{\sum_N (PM_n - MM_n)\phi_n}{N}$$

where ϕ_n is a scaling factor that is specific to probe pair $PM_n - MM_n$ and is obtained by fitting a statistical model to a series of experiments. This model takes into account that probe pairs respond differently to changes in expression of a gene and that the variation between replicates is also probe-pair dependent. This probe-specific analysis can also be used as an alternative to mismatch probes, because any nonspecific binding or crosshybridization—usually revealed by mismatch probes—should also be revealed by the behavior of perfect match probes across experiments. Software for fitting the model (weighted average difference and weighted perfect match), as well as for detecting outliers and obtaining estimates on reliability is available for download.[2] Lemon and coworkers have compared the Li-Wong model to Affymetrix' Average Difference and found it to be superior in a realistic experimental setting (Lemon, et al., 2001)

3.2 SCALING

The aim is to identify genes that are up- or down-regulated in the sample compared to the control. How do you know that you can compare the sample and the control? You don't unless you have *scaled* them. You need to ensure that the expression levels in the sample are comparable to the expression levels in the control. You can do that by including so-called maintenance genes on the chip, genes that are assumed to be constitutively expressed and expressed at a constant level. Then you check that these genes are expressed at the same level in the control and in the sample. If they are not, you multiply all expression values in the sample by some factor until the maintenance genes match the expression levels in the control. It is rare, however, that maintenance genes are completely unaffected by your experiment.

Another way to scale the chips is to assume that the total amount of mRNA measured from each cell is constant. This will obviously not hold true if you

---

[1] http://www.affymetrix.com/products/statistical_algorithms_reference_guide.html
[2] http://www.dchip.org

have spotted a chip with only two genes. But the more genes there are on your chip, the more realistic the assumption of constant total mRNA will be. You then multiply all expression levels by a scaling factor until the sum of all expression levels in the sample is equal to the sum of all expression levels in the control (or you set a target average expression for both).

Affymetrix in the early versions of their software used this scaling, except that the highest 2% and lowest 2% Average Difference values were excluded from the calculation of average expression.

Before scaling you should make sure that all your measurements are within the linear range of your equipment. Otherwise the scaling will introduce errors that will bias any comparison between two chips. Affymetrix equipment in early experiments had a tendency to reach saturation on highly expressed genes if antibody amplification of the cRNA (*in vitro* transcribed cDNA) was employed. If you plotted unamplified Affymetrix data against amplified data you would see the saturation as well as other sources of nonlinearity. When you scale on nonlinear data you introduce artifacts.

Such nonlinear data are better scaled by scaling the weakly and the highly expressed genes separately. Within a small range there is linearity and a simple scaling factor can be used (see also Schadt, et al., 2000). Li and Wong (2001b) used a piecewise linear running median to scale chip data.[3] In our lab, we have developed (Workman, et al., 2001) qspline, a series of tools that fit a cubic spline (a polynomium of the third degree) to the cumulative distribution of intensities.[4]

Our software also handles the saturation that was seen with early antibody-amplified Affymetrix chips. It substitutes intensities from saturated probes with intensities from the same probes measured on the unamplified chips—after proper scaling, of course.

A special situation arises when a large proportion of genes is affected in the experiment. Global effects such as starvation or heat shock may alter the expression of so many genes that a global normalization will obscure the true regulation. In that case another alternative is available. You can use spike (hybridization) controls, addition of messengers from a foreign organism for which you have probes on the chip. If you add spike controls in equal amounts to mRNA preparations, you can use their intensity to scale. But how do you know that the mRNA preparations are comparable at the time you add the spike control to them? Use spike control scaling only if you have no other choice.

---

[3] Software available at http://www.dchip.org
[4] Software available at http://www.cbs.dtu.dk/biotools/oligoarray/

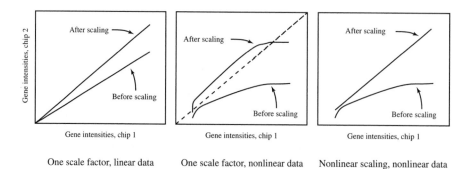

Fig. 3.1   Illustration of scaling techniques.

3.2.1   Example: Linear and Nonlinear Scaling

The scaling with one scale factor works well for linear data with zero y-intercept as shown in Figure 3.1 (left). The intensities for each gene measured on two chips is plotted on the x and y axis. Before scaling the genes do not lie along the diagonal, meaning that scaling is necessary to make the chips comparable. If there is a linear relationship between the intensities of the two chips (left graph) and a zero y-intercept, a single scale factor will do the job. After all genes on chip 1 have been multiplied by this factor, most of them match the intensity measured in chip 2.

For nonlinear, nonzero y-intercept data (Figure 3.1 center graph), however, this method fails. Such nonlinearities can be observed in early Affymetrix data, for example, if a chip is measured before and after amplification and the two scans are compared to each other. After scaling chip 1 with a single factor, all genes are off the diagonal, implying different expression on the two chips when it should be the same for most genes.

For such nonlinear data, applying different scaling factors to different segments of the curve or using a nonlinear fit to the data is the right way to go (Figure 3.1 right graph).

So how many scale factors (or how many parameters in a nonlinear scaling) should you use? It is obvious that if you use as many scale factors as there are genes you scale out the signal, those genes that are significantly different between the two chips, as well. Schadt et al. (2000) use two scale factors.

3.3   DETECTION OF OUTLIERS

Outliers in chip experiments can occur at several levels. You can have an entire chip that is bad and consistently deviates from other chips made from the same condition or sample. Or you can have an individual gene on a chip that deviates from the same gene on other chips from the same sample. That can be caused by image artifacts such as hairs, air bubbles, precipitation and so on. Finally, it is possible that a single probe, due to precipitation or other artifact, is perturbed.

How can you detect outliers in order to remove them? Basically, you need a statistical model of your data. The simplest model is equality among replicates. If one replicate (chip, gene, or probe) deviates several standard deviations from the mean, you can consider it an oulier and remove it. The *t*-test measures standard deviation and gives genes where outliers are present among replicates a low significance (See Section 3.5).

More advanced statistical models have been developed that also allow for outlier detection and removal (Li and Wong, 2001a, b).

3.4 FOLD CHANGE

Having performed this scaling, you should now be able to compare the expression level of any gene in the sample to the expression level of the same gene in the control. The next thing you want to know is, How many fold up- or down-regulated is the gene, or is it unchanged?

To calculate fold change the simplest approach is to divide the expression level of a gene in the sample by the expression level of the same gene in the control. Then you get the fold change, which is 1 for an unchanged expression, less than 1 for a down-regulated gene, and larger than 1 for an up-regulated gene. The definition of fold change will not make any sense if the expression value in the sample or in the control is zero or negative. Early Average Difference values from Affymetrix sometimes were, and a quick-and-dirty way out of this problem was to set all Average Difference values below 20 to 20. This was the approach usually applied.

The problem with fold change emerges when one takes a look at a scale. Up-regulated genes occupy the scale from 1 to infinity (or at least 1000 for a 1000-fold up-regulated gene) whereas all down-regulated genes only occupy the scale from 0 (0.001 for a 1000-fold down-regulated gene) to 1. This scale is highly asymmetric.

The Affymetrix GeneChip software (early versions) calculates fold change in a slightly different way that does stretch out that scale to be symmetric:

$$AffyFold = \frac{Sample - Control}{min(Sample, Control)} + \begin{cases} +1 & \text{if Sample} > \text{Control} \\ -1 & \text{if Sample} < \text{Control} \end{cases}$$

where *Sample* and *Control* are the AvgDiffs of the sample and the control, repectively. For calculation of fold change close to the background level, consult the Affymetrix manual.

This function is discontinuous and has no values in the interval from -1 to 1. Up-regulated genes have a fold change greater than 1 and down-regulated genes have a fold change less than -1. But the scale for down-regulated

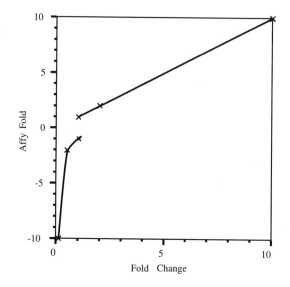

Fig. 3.2   AffyFold vs. fold change. Points show no change, 2-, and 10-fold up- and down-regulation.

genes is comparable to the scale for up-regulated genes (see Figure 3.2); the scale is symmetric around 1.

Both the fold change and Affyfold expressions are intuitively rather easy to grasp and deal with, but for further computational data analysis they are not useful, either because they are asymmetric or because they are discontinuous. For further data analysis you need to calculate the logarithm of fold change (Figure 3.3). Logfold, as we will abbreviate it, is undefined for the Affymetrix fold change, but can be applied to the simple fold change provided that you have taken the precaution to avoid values with zero or negative expression.

It is not important whether you use the natural logarithm ($\log_e$), base 2 logarithm ($\log_2$), or base 10 logarithm ($\log_{10}$), as long as you are consistent!

3.5   SIGNIFICANCE

If you have found a gene that is twofold up-regulated ($\log_{10}$ fold is 0.3), then how do you know whether this is just a result of experimental error? You need to determine whether or not a twofold regulation is *significant*. There are several ways to estimate significance in chip experiments, and new methods are being developed. Basically, to assess experimental error you have to repeat the experiment and measure the variation. If you repeat both control and sample, you can use a *t*-test to determine whether the expression of a particular gene is significantly different between control and sample. (The

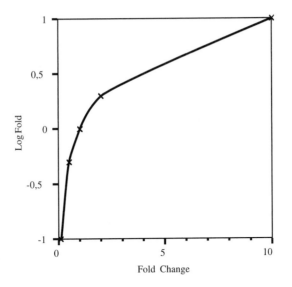

Fig. 3.3   Logfold vs. fold change. Points show no change, 2-, and 10-fold up- and down-regulation.

t-test looks at the mean and variance of the two distributions and calculates the probability that they were sampled from the same distribution. If you subtract that probability from 1, you get the probability that they were sampled from different distributions.)   When using the *t*-test it is important not to assume equal variance between sample and control.   The readings from the sample experiments could be $1120, 1320, 980$ and the readings from the control experiment could be $5, 35, 40$.   Here you clearly have much larger variance in the sample than in the control.   Welch's *t*-test takes this into account by assuming unequal variances between the two populations.

If you have repeat measurements only for the sample and not for the control, you can estimate experimental error from the variation between replicates. You can determine the standard deviation for each gene and then compare the change in expression to the standard deviation (Tusher, et al., 2001). The more the change exceeds the standard deviation between replicates, the more significant it is. But it does not allow you to calculate a *P*-value. Instead you can estimate the false positive rate through permutation of the data (Tusher, et al., 2001).

If you compare sample and control experiment without any replication at all, you run the risk of identifying the most unstable messengers as differentially expressed. The messengers that are most unstable may be degraded during sampling and/or sample preparation, and this degradation is very sensitive to small details of your extraction procedure. Thus you may have more degradation in the sample than in the control, leading you to falsely conclude that the messenger is down-regulated.

The only way to avoid making this serious mistake is to replicate the experiment and use the *t*-test to look for messengers that reproducibly are changed in concentration.

If you have more than two conditions, the *t*-test is not the method of choice. The method analysis of variance (ANOVA) will, using the F distribution, calculate the probablity that several conditions, each replicated, all come from the same distribution. If you subtract that probability from 1, you get the probability that a gene has a significantly changed expression in at least one condition.

Software for running the *t*-test and ANOVA will be discussed in Section 12.4. A web-based method for *t*-test is available[5] (Baldi and Long, 2001). Baldi and Long (2001) advocate using the *t*-test on log-transformed data.

3.5.1 Nonparametric Tests

Both the *t*-test and ANOVA assume that your data follow the normal distribution. Although both methods are robust to moderate deviations from the normal distribution, alternative methods exist for assessing significance without assuming normality. The Wilcoxon/Mann-Whitney rank sum test will do the same without using the actual expression values from the experiment, only their rank relative to each other.

When you rank all expression levels from the two conditions, the best separation you can have is that all values from one condition rank higher than all values from the other condition. This corresponds to two nonoverlapping distributions in parametric tests. But since the Wilcoxon test does not measure variance, the significance of this result is limited only by the number of replicates in the two conditions. It is for this reason that you may find that the Wilcoxon test for low numbers of replication gives you a poor significance and that the distribution of *P*-values is rather granular.

3.5.2 Correction for Multiple Testing

For all statistical tests that calculate a *P*-value, it is important to consider the effect of multiple testing as we are looking at not just one gene but thousands of genes. If a *P*-value of 0.01 tells you that you have a probablity of 1% of being wrong on one gene, then you expect one false positive gene if you look at 100 genes. If you look at 7000 genes, you expect 70 false positives. That is not what you want! If you want a chance of, say, 25% of having

---

[5]http://128.200.5.223/CyberT/

Table 3.1 Expression readings of four genes in six patients.

| Gene | Patient | | | | | |
	N_1	N_2	A_1	A_2	B_1	B_2
a	90	110	190	210	290	310
b	190	210	390	410	590	610
c	90	110	110	90	120	80
d	200	100	400	90	600	200

one false positive in your list of genes that show significant regulation in an experiment, you need to perform a Bonferroni correction (Dudoit, et al., 2000; Bender and Lange, 2001): 0.25 divided by 7000 genes gives a P-value cutoff of $3 \cdot 10^{-5}$. If you only consider P-values more significant than this cutoff then your chance of having any false positives on the list is only 25%. This is a very conservative interpretation of your results but it is a simple and efficient guard against false positive conclusions.

If no genes in your experiment pass this very conservative Bonferroni correction, then you can look at those that have the smallest P-value. When you multiply their P-value by the number of genes in your experiment, you get an estimate of the number of false positives. Take this false positive rate into account when planning further experiments.

3.5.3 Example I: *t*-Test and ANOVA

A small example using only four genes will illustrate the *t*-test and ANOVA. The four genes are each measured in six patients, which fall into three categories: normal (N), disease stage A, and disease stage B. That means that each category has been *replicated* once (Table 3.1).

We can perform a *t*-test (see Section 12.4 for details) to see if genes differ significantly between patient category A and patient category B (Table 3.2).

Table 3.2 *t*-test on difference between patient categories A and B.

| Gene | Patient | | | | |
	A_1	A_2	B_1	B_2	P-value
a	190	210	290	310	0.019
b	390	410	590	610	0.005
c	110	90	120	80	1.000
d	400	90	600	200	0.606

Table 3.3 ANOVA on difference between patient categories N, A and B.

Gene	N_1	N_2	A_1	A_2	B_1	B_2	P-value
a	90	110	190	210	290	310	0.0018
b	190	210	390	410	590	610	0.0002
c	90	110	110	90	120	80	1.0000
d	200	100	400	90	600	200	0.5560

But you should be careful performing a t-test on as little as two replicates in real life. This is just for illustration purposes.

Gene b is significantly different at a 0.05 level, even after multiplying the P-value by four to correct for multiple testing. Gene a is not significant at a 0.05 level after Bonferroni correction, and genes c and d have a high probability of being unchanged. For gene d that is because, even though an increasing trend is observed, the variation within each category is too high to allow any conclusions.

If we perform an ANOVA instead, testing for genes that are significantly different in at least one of three categories, the picture changes slightly (Table 3.3).

In the ANOVA, both gene a and b are significant at a 0.01 level even after Bonferroni correction. So taking all three categories into account increases the power of the test relative to the t-test on just two categories.

3.5.4   Example II: Number of Replicates

If replication is required to determine the significance of results, how many replicates are required? An example will illustrate the effect of the number of replicates. We have performed six replicates of each of two conditions in a *Saccharomyces cerevisiae* GeneChip experiment (Piper, et al., 2002). Some of the replicates have even performed in different labs. Accepting the results of a t-test on this data set as the correct answer, we can ask, How close would five replicates have come to that answer? How close would four replicates have come to that answer? We have performed this test (Piper, et al., 2002). and Table 3.4 shows the results. For each choice of replicates, we show how many false positives (Type I errors) we have relative to the correct answer and how many false negatives (Type II errors) we have relative to the correct answer. The number of false positives in the table is close to the number we have chosen with our cutoff in the Bonferroni corrected t-test (a 0.005 cutoff at 6383 genes yields 32 expected false positives). The number of false negatives, however, is greatly affected by the number of replicates.

Table 3.4   Effect of number of replicates on Type I (FP) and II (FN) errors in *t*-test.

	Number of replicates of each condition				
	2	3	4	5	6
True positives	23	144	405	735	1058
False positives	8	18	29	45	0
False negatives	1035	914	653	323	0

Table 3.5   Effect of number of replicates on Type I and II errors in SAM (Tusher, et al., 2001).

	Number of replicates of each condition				
	2	3	4	5	6
True positives	27	165	428	748	1098
False positives	3	4	14	27	0
False negatives	1071	933	670	350	0

Table 3.5 shows that the *t*-test performs almost as well as SAM[6] (Tusher, et al., 2001), which has been developed specifically for estimating the false positive rate in DNA microarray experiments based on permutations of the data.

While this experiment may not be representative, it does illustrate two important points about the *t*-test. You can control the number of false positives even with very low numbers of replication. But you lose control over the false negatives as the number of replications go down.

So how many replicates do you have to perform to avoid any false negatives? That depends mainly on two parameters. How large is the variance between replicates and how small a fold change do you wish to detect. Both parameters are gene specific, so it is not possible to give a general answer to the number of replicates required. Again our experiment will serve as an illustration of how these two parameters are related.

Using the power.t.test function of the R statistics software package (see Section 12.4) we have calculated for each gene the minimum fold change required to be significant at a 0.005 level and with a 0.2 probability of false negatives. That depends on the variance, which we calculate for each gene with the six replicates, and the mean expression level, which we calculate for each gene as the mean of the six replicates

---

[6]Software available for download at http://www-stat.stanford.edu/~tibs/SAM/index.html

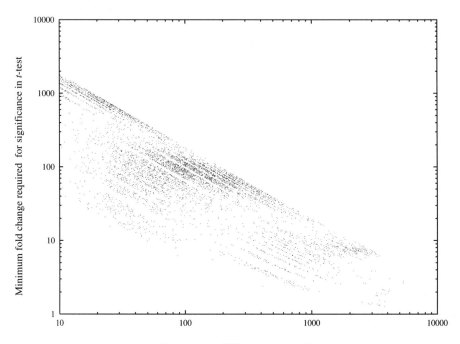

Fig. 3.4   Minimum fold change required for significant change call. Based on six replicates of a single condition, the minimum fold change for a 0.005 significance for each gene is shown as a function of its mean Average Difference value in the six replicates (genes with mean Average Difference below 10 are omitted).

From Figure 3.4 it is evident that there are highly expressed genes that have such a low variance between replicates that a fold change less than two can be determined with significance. But it is also evident that, for more weakly expressed genes with high variance between replicates, a fold change of up to thousand may be required for significant detection of change.

3.6   SUMMARY

Whether you have intensities from a spotted array or Average Difference (use Li and Wong's weighted average difference, if possible) from an Affymetrix chip, the following suggestions apply:

- The standard scaling with one factor to get the same average intensity in all chips is a good way to start, but it is not the best way to do it. Use scaling with multiple factors or polynomial fitting if possible.

- Repeat each condition of the experiment (as a rule-of-thumb at least three times) and apply a statistical test for significance of observed differences. Apply the test on the scaled intensities (or AvgDiff). For spotted arrays with large variation between slides you can consider applying the statistical test on the fold change from each slide as well.

- Correct the statistical test for multiple testing (Bonferroni correction or similar).

3.7 FURTHER READING

Vingron, M. (2001). Bioinformatics Needs to adopt statistical thinking (Editorial). *Bioinformatics* 17:389–390.

Student's t-test, ANOVA, and Wilcoxon/Mann-Whitney

Kerr, M. K., Martin, M., and Churchill, G. A. (2000). Analysis of variance for gene expression microarray data. *J. Comput. Biol.* 7:819–37.

Kerr, M. K., and Churchill, G. A. (2001). Statistical design and the analysis of gene expression microarray data. *Genet Res.* 77:123–8. Review.

Montgomery, D. C., and Runger, G. C. (1999). *Applied Statistics and Probability for Engineers.* New York: Wiley.

Number of replicates

Pan, W., Lin, J., and Le, C. (2001). How many replicates of arrays are required to detect gene expression changes in microarray experiments? A mixture model approach. Report 2001-012, Division of Biostatistics, University of Minnesota.[7]

Piper, M., Daran-Lapujade, P., Bro, C., Regenberg, B., Knudsen, S., Nielsen, J., and Pronk, J. (2002). Reproducibility of transcriptome analyses using oligonucleotide microarrays: An interlaboratory comparison of chemostat cultures of *Saccharomyces cerevisiae*. Manuscript in preparation.

---

[7] Available at http://www.biostat.umn.edu/cgi-bin/rrs?print+2001

Correction for multiple testing

Dudoit, S., Yang, Y., Callow, M. J., and Speed, T. P. (2000). Statistical methods for identifying differentially expressed genes in replicated cDNA microarray experiments. Technical report #578, August 2000.[8]

Bender, R., and Lange, S. (2001). Adjusting for multiple testing—when and how? *Journal of Clinical Epidemiology* 54:343–349.

A Bayesian expansion of the t-test that estimates variance by assuming that genes with similar expression level have similar variance

Baldi, P., and Long, A. D. (2001). A Bayesian framework for the analysis of microarray expression data: Regularized t-test and statistical inferences of gene changes. *Bioinformatics* 17:509–519.[9]

Average Difference calculation and outlier detection

Lemon, W. J., Palatini, J. T., Krahe, R., and Wright, F. A. (2001). Comparison of gene expression estimators for oligonucleotide arrays. Submitted 2001.[10]

Li, C., and Wong, W. H. (2001a). Model-based analysis of oligonucleotide arrays: Expression index computation and outlier detection. *Proc. Natl. Acad. Sci. USA* 98:31–36.[11]

Nonparametric significance tests developed for array data

Efron, B., Storey, J., and Tibshirani, R. (2001). Microarrays, empirical Bayes methods, and false discovery rates. Technical report. Statistics Department, Stanford University.[12]

Park, P. J., Pagano, M., and Bonetti, M. (2001). A nonparametric scoring algorithm for identifying informative genes from microarray Data. *Pacific Symposium on Biocomputing* 6:52–63.[13]

Other significance tests developed for array data

Ideker, T., Thorsson, V., Siegel, A. F., and Hood, L. (2000). Testing for differentially-expressed genes by maximum-likelihood analysis of microarray data. *Journal of Computational Biology* 7:805–817.

---

[8] Available at http://www.stat.berkeley.edu/tech-reports/index.html
[9] Accompanying web page at http://128.200.5.223/CyberT/
[10] Preprint and Perl scripts available at http://thinker.med.ohio-state.edu/projects/fbss/index.html
[11] Software available at http://www.dchip.org
[12] Manuscript available at http://www-stat.stanford.edu/~tibs/research.html
[13] Manuscript available online at http://psb.stanford.edu

Newton, M. A., Kendziorski, C. M., Richmond, C. S., Blattner, F. R., and Tsui, K. W. (2001). On differential variability of expression ratios: Improving statistical inference about gene expression changes from microarray data. *J. Comput. Biol.* 8:37–52.

Thomas, J. G., Olson, J. M., Tapscott, S. J., and Zhao, L. P. (2001). An efficient and robust statistical modeling approach to discover differentially expressed genes using genomic expression profiles. *Genome Res.* 11:1227–1236.

Tusher, V. G., Tibshirani, R., and Chu, G. (2001). Significance analysis of microarrays applied to the ionizing radiation response. *Proc. Natl. Acad. Sci. USA* 98:5119–5121.[14]

Zhao, L. P., Prentice, R., and Breeden, L. (2001). Statistical modeling of large microarray data sets to identify stimulus-response profiles. *Proc. Natl. Acad. Sci. USA* 98:5631–5636.

Scaling or normalization

Dudoit, S., Yang, Y., Callow, M. J., and Speed, T. P. (2000). Statistical methods for identifying differentially expressed genes in replicated cDNA microarray experiments Technical report #578, August 2000.[15]

Goryachev, A. B., Macgregor, P. F., and Edwards, A. M. (2001). Unfolding of microarray data. *Journal of Computational Biology* 8:443–61.

Li, C., and Wong, W. H. (2001b). Model-based analysis of oligonucleotide arrays: Model validation, design issues and standard error application. *Genome Biology* 2:1–11.[16]

Schadt, E. E., Li, C., Su, C., and Wong, W. H. (2000). Analyzing high-density oligonucleotide gene expression array data. *J. Cell. BioChem.* 80:192–201.

Schuchhardt, J., Beule, D., Malik, A., Wolski, E., Eickhoff, H., Lehrach, H., and Herzel, H. (2000). Normalization strategies for cDNA microarrays. *Nucleic Acids Res.* 28:E47.

Workman, C., Jensen, L. J., Jarmer, H., Saxild, H. H., Berka, R., Gautier, L., Nielsen, H. B., Nielsen, C., Brunak, S., and Knudsen, S. (2001). A new non-linear method to reduce variability in DNA microarray experiments. Manuscript submitted.[17]

---

[14]Software available for download at http://www-stat.stanford.edu/~tibs/SAM/index.html

[15]Available at http://www.stat.berkeley.edu/tech-reports/index.html

[16]Software available at http://www.dchip.org

[17]Software available at http://www.cbs.dtu.dk/biotools/oligoarray/

Zien, A., Aigner, T., Zimmer, R., and Lengauer, T. (2001). Centralization: A new method for the normalization of gene expression data. *Bioinformatics* 17(Suppl 1):S323–S331.

4

Visualization by Reduction of Dimensionality

The data from expression arrays is often of high dimensionality. If you have measured 6000 genes in 15 patients, the data constitute a matrix of 15 by 6000. It is impossible to discern any trends by visual inspection of such a matrix. It is necessary to reduce the dimensionality of this matrix to allow visual analysis. Since visual analysis is traditionally performed in two dimensions, in a coordinate system of x and y, many methods allow reduction of a matrix of any dimensionality to just two dimensions. These methods include principal component analysis, correspondence analysis, singular value decomposition, multidimensional scaling, and cluster analysis.

4.1 PRINCIPAL COMPONENT ANALYSIS

If we want to display the data in just two dimensions, we want as much of the variation in the data as possible captured in just these two dimensions. Principal component analysis (PCA) has been developed for this purpose. Imagine 6000 genes as points in hyperspace, each dimension corresponding to its expression in one of 15 patients. You will see a cloud of 6000 points in hyperspace. But the cloud is not hyperspherical. There will be one direction in which the cloud will be more irregular or extended. (Figure 4.1 illustrates this with only a few points in three dimensions.) This is the axis of the first principal component. This axis will not necessarily coincide with one of the patient axes. Rather, it will have projections of several patient axes on it. Next, we look for an axis that is orthogonal to the first principal component,

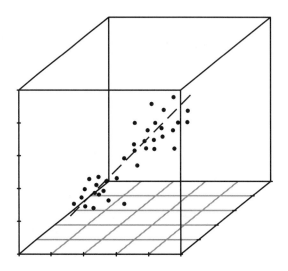

Fig. 4.1 A cloud of points in three-dimensional space. The cloud is not regular. It extends more in one direction than in all other directions. This direction is the first principal component (dashed line).

and captures the maximum amount of variation left in the data. This is the second principal component. We can now plot all 6000 genes in these two dimensions. We have reduced the dimensionality from 15 to 2, while trying to capture as much variation in the data as possible. The two principal components have been constructed as sums of the individual patient axes.

What will this analysis tell you? Perhaps nothing; it depends on whether there is a trend in your data that is discernible in two dimensions. Other relationships can be visualized with cluster analysis, which will be described in Section 5.

Instead of reducing the patient dimensions we can reduce the gene dimensions. Why not throw out all those genes that show no variation? We can achieve this by performing a principal component analysis of the genes. We are now imagining our data as 15 points in a space of 6000 dimensions, where each dimension records the expression level of one gene. Some dimensions contribute more to the variation between patients than other dimensions. The first principal component is the axis that captures most variation between patients. A number of genes have a projection on this axis, and the principal component method can tell you how much each gene contributes to the axis. The genes that contribute the most show the most variation between patients. Can this method be used for selecting diagnostic genes? Yes, but it is not necessarily the best method, for variation, as we have seen in Section 3.5, can be due to noise as well as true difference in expression. Besides, how do we know how many genes to pick? The *t*-test and ANOVA, mentioned in

Table 4.1 Expression readings of four genes in six patients.

Gene	N_1	N_2	A_1	A_2	B_1	B_2
			Patient			
a	90	110	190	210	290	310
b	190	210	390	410	590	610
c	90	110	110	90	120	80
d	200	100	400	90	600	200

Section 3.5, are more suited to this task. They will, based on a replication of measurement in each patient, tell you which genes vary between patients and give you the probability of false positives at the cutoff you choose.

So a more useful application of principal component analysis would be to visualize genes that have been found by a t-test or ANOVA to be significantly regulated. This visualization may give you ideas for further analysis of the data.

4.2 EXAMPLE 1: PCA ON SMALL DATA MATRIX

Let us look at a simple example to visualize the problem. We have the data matrix shown in Table 4.1.

It consists of four genes measured in six patients. If we perform a principal component analysis on this data (the details of the computation are shown in Section 12.4), we get the biplot shown in Figure 4.2. A biplot is a plot designed to visualize both points and axes simultaneously. Here we have plotted the four genes as points in two dimensions, the first two principal components. It can be seen that genes a, c, and b differ a lot in the first dimension (they vary from about -400 to $+400$) while they differ little in the second dimension. Gene d, however, is separated from the other genes in the second dimension (it has a value of about -400 in the second dimension).

Indicated as arrows are the projections of the six patient axes on the two first principal components. Start with Patient B_1. This patient has a large projection (about 0.5) on the first principal component, and a smaller projection on the second principal component (about -0.3). The lengths of the patient vectors indicate how much they contribute to each axis and their directions indicate in which way they contribute. The first principal component consists mainly of Patient category B, where expression differs most. Going back to the genes, it can be seen that they are ranked according to average expression level in the B patients along this first principal component: genes c, a, d and b.

The second principal component divides genes into those that are higher in Patient B_2 than in Patient B_1 (gene c, a, and b), and gene d, which is lower

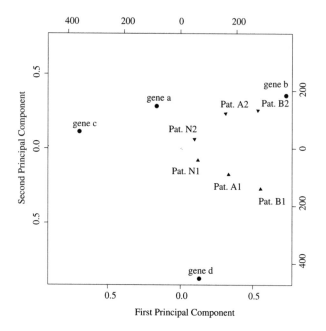

Fig. 4.2 Principal Component Analysis of data shown in Table 4.1.

in Patient B_2 than in Patient B_1. On the vector projections of the patient axes on this component it can be seen that they have been divided into those with subcategory 1 (Patients N_1, A_1, B_1), which all have a positive projection, and those with subcategory 2 (Patients N_2, A_2, B_2), which all have a negative projection. So the second principal component simply compares expression in subcategories.

We can also do the principal component analysis on the reverse (*transposed*) matrix (transposition means to swap rows with columns). Figure 4.3 shows a biplot of patients along principal components that consist of those genes that vary most between patients. First, it can be seen that there has been some grouping of patients into categories. Categories can be separated by two parallel lines. By looking at the projection of the gene vectors we can see that gene b and gene d, those that vary most, contribute most to the two axes. Now, if we wanted to use this principal component analysis to select genes that are diagnostic for the three categories, we might be tempted to select gene b and gene d because they contribute most to the first principal component. This would be a mistake, however, because gene d just shows high variance that is not correlated to category at all. The ANOVA, described in Section 3.5, would have told us that gene b and gene a are the right genes to pick as diagnostic genes for the disease.

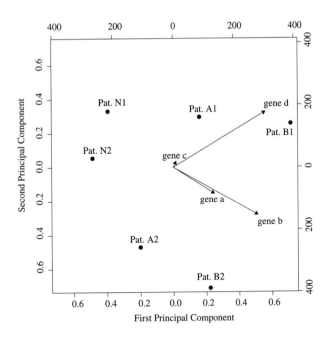

Fig. 4.3   Principal Component Analysis of transposed data of Table 4.1.

4.3   EXAMPLE 2: PCA ON REAL DATA

Figure 4.4 shows results of a PCA on real data. The R package was used on the HIV data (Section 1.5) as described in Section 12.4. The projections of the 7 experiments (4 controls (C) and 3 HIV (H)) on the principal components are shown as vectors in this biplot. The first principal component captures overall differences in expression level among genes—it separates them into those with negative expression (AvgDiff) and those with high expression. The second principal component separates the genes into those expressed higher in HIV than in the controls and those expressed higher in the controls than in HIV. What is ignored in this separation, however, is the variance between replicates in each group. The t-test (Section 3.5) makes a better selection of differentially expressed genes because it takes into account variance between replicates.

4.4   SUMMARY

Principal component analysis is a way to reduce your multidimensional data to a single $x - y$ graph. You may be able to spot important trends in your

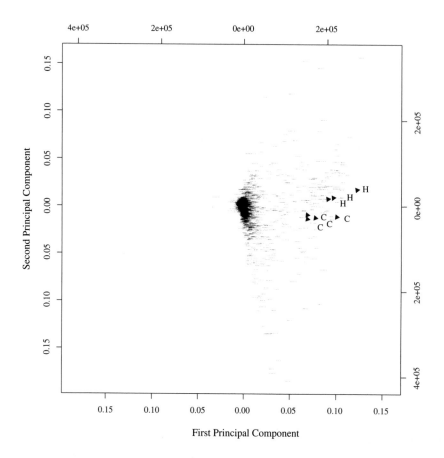

Fig. 4.4 Principal component analysis on real data from HIV experiment. All genes are plotted along the first two principal components. Genes are indicated by their name, but there are too many and the font is too small to be legible in this plot. The projections of the 7 experiments (4 controls (C) and 3 HIV (H)) on the principal components are shown as vectors in this biplot.

data from this one graph alone. If replicates are available it is best to perform PCA on data that has already been filtered for significance.

4.5 FURTHER READING

Singular Value Decomposition

Alter, O., Brown, P. O., and Botstein, D. (2000). Singular value decomposition for genome-wide expression data processing and modeling. *Proc. Natl. Acad. Sci. USA* 97:10101–6.

Holter, N. S., Mitra, M., Maritan, A., Cieplak, M., Banavar, J. R., and Fedoroff, N.V. (2000). Fundamental patterns underlying gene expression profiles: Simplicity from complexity. *Proc. Natl. Acad. Sci. USA* 97:8409–14.

Wall, M. E., Dyck, P. A., and Brettin, T. S. (2001). SVDMAN—singular value decomposition analysis of microarray data. *Bioinformatics* 17:566–568.

Principal component analysis

Dysvik, B, and Jonassen, I. (2001). J-Express: Exploring gene expression data using Java. *Bioinformatics* 17:369–70.[1]

Raychaudhuri, S., Stuart, J. M., and Altman, R. B. (2000). Principal components analysis to summarize microarray experiments: Application to sporulation time series. *Pac. Symp. Biocomput.* 2000:455–66.[2]

Xia, X, and Xie, Z. (2001). AMADA: Analysis of microarray data. *Bioinformatics* 17:569–70.

Xiong, M., Jin, L., Li, W., and Boerwinkle, E. (2000). Computational methods for gene expression-based tumor classification. *Biotechniques* 29:1264–1268.

Correspondence analysis

Fellenberg, K., Hauser, N. C., Brors, B., Neutzner, A., Hoheisel, J. D., and Vingron, M. (2001), Correspondence analysis applied to microarray data. *Proc. Natl. Acad. Sci. USA* 98:10781–10786.

Gene Shaving uses PCA to select genes with maximum variance

Hastie, T., Tibshirani, R., Eisen, M. B., Alizadeh, A., Levy, R., Staudt, L., Chan, W. C., Botstein, D., and Brown, P. (2000). Gene shaving as a method for identifying distinct sets of genes with similar expression patterns. *Genome Biol.* 1:RESEARCH0003.1–21

---

[1] Software available at http://www.ii.uib.no/~bjarted/jexpress/
[2] Available online at http://psb.stanford.edu

5

Cluster Analysis

If you have just one experiment and a control, your first data analysis will limit itself to a list of regulated genes ranked by the magnitude of up- and down-regulation, or ranked by the significance of regulation determined in a *t*-test.

Once you have more experiments—measuring the same genes under different conditions, in different mutants, in different patients, or at different time points during an experiment—it makes sense to group the significantly changed genes into clusters that behave similarly under the different conditions.

5.1 HIERARCHICAL CLUSTERING

Think of each gene as a vector of N numbers, where N is the number of experiments or patients. Then you can plot each gene as a point in N-dimensional hyperspace. You can then calculate the distance between two genes as the Euclidean distance between their respective data points (as the root of the sum of the squared distances in each dimension).

This can be visualized using a modified version of the small example data set applied in previous chapters (Table 5.1). The measured expression level of the five genes can be plotted in just two of the patients using a standard $x - y$ coordinate system (Figure 5.1 left).

You can calculate the distance between all genes (producing a *distance matrix*), and then it makes sense to group those genes together that are closest

Table 5.1   Expression readings of five genes in two patients.

| Gene | Patient | |
	N_1	A_1
a	90	190
b	190	390
c	90	110
d	200	400
e	150	200

to each other in space. The two genes that are closest to each other, b and d, form the first cluster (Figure 5.1 left). Genes a and e are separated by a larger distance, and they form a cluster as well (Figure 5.1 left). If the separation between a gene and a cluster comes within the distance as you increase it, you add that gene to the cluster. Gene c is added to the cluster formed by a and e. How do you calculate the distance between a point (gene) and a cluster? You can calculate the distance to the *nearest neighbor* in the cluster (gene e), but it is more unbiased to calculate the distance to the point that is in the middle of the existing members of the cluster (centroid, similar to UPGMA or average linkage method).

When you have increased the distance to a level where all genes fall within that distance, you are finished with the clustering and can connect the final clusters. You have now performed a *hierarchical agglomerative clustering*. There are computer algorithms available for doing this (see Section 12.3).

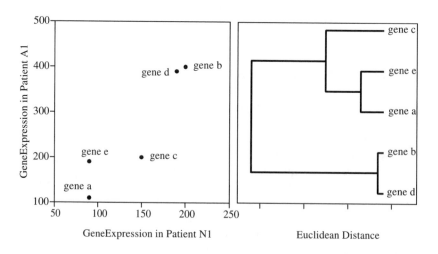

Fig. 5.1   Hierarchical clustering of genes based on their Euclidean distance.

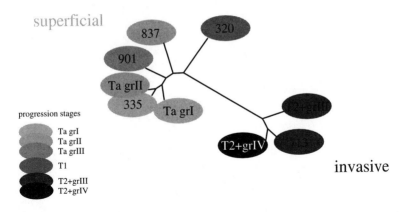

Fig. 5.2 Hierarchical clustering of bladder cancer patients. The clustering was based on expression measurements from a DNA array hybridized with mRNA extracted from a biopsy. (Christopher Workman, based on data published in Thykjaer, et al. (2001)). (See color plate.)

A real example is shown in Figure 5.2 where bladder cancer patients were clustered based on Affymetrix GeneChip expression measurements from a bladder biopsy. It is seen in the figure that the clustering groups superficial tumors together and groups invasive tumors together.

Hierarchical clustering only fails when you have a large number of genes (several thousand). Calculating the distances between all of them becomes time consuming. Removing genes that show no significant change in any experiment is one way to reduce the problem. Another way is to use a faster algorithm, like *K*-means clustering.

5.2 *K*-MEANS CLUSTERING

In *K*-means clustering, you skip the calculation of distances between all genes. You decide on the number of clusters you want to divide the genes into, and the computer then randomly assigns each gene to one of the *K* clusters. Now it will be comparatively fast to calculate the distance between each gene and the center of each cluster (*centroid*). If a gene is actually closer to the center of another cluster than the one it is currently assigned to, it is reassigned to the closer cluster. After assigning all genes to the closest cluster, the centroids are recalculated. After a number of iterations of this, the cluster centroids will no longer change, and the algorithm stops. This is a very fast algorithm, but it will give you only the number of clusters you asked for and not show their relation to each other as a full hierarchical clustering will do. In practice, *K*-means is useful if you try different values of *K*.

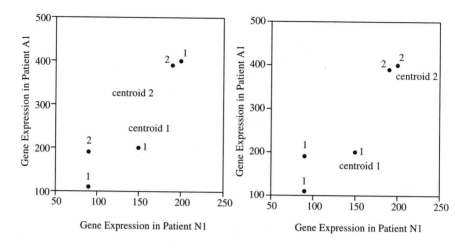

Fig. 5.3 *K*-means clustering of genes based on their Euclidean distance. First, genes are randomly assigned to one of the two clusters in *K*: 1 or 2 (Left). The centroids of each cluster are calculated. Genes are then reassigned to another cluster if they are closer to the centroid of that cluster (Right). After just one iteration, the final solution is obtained (Right).

If you try the *K*-means clustering on the expression data used for hierarchical clustering shown in Figure 5.1, with *K* = 2, the algorithm may find the solution in just one iteration (Figure 5.3).

5.3 SELF-ORGANIZING MAPS

There are other methods for clustering, but hierarchical and *K*-means cover most needs. One method that is available in a number of clustering software packages is self-organizing maps (SOM) (Kohonen, 1995). SOM is similar to *K*-means, but instead of allowing centroids to move freely in (multidimensional) space, they are constrained to a two-dimensional grid. The algorithm then organizes itself to best accommodate the data in this grid. The end result is a clustering of the data in a grid, two-by-two, four-by-four, or whatever size is specified. The grid structure implies relationships between neighboring clusters on the grid.

Figure 5.4 shows how a SOM clustering could fit a two-by-two grid on the data of our example (the cluster centers are at each corner of the grid). That would result in four clusters, three of them with one member only and one cluster with two members.

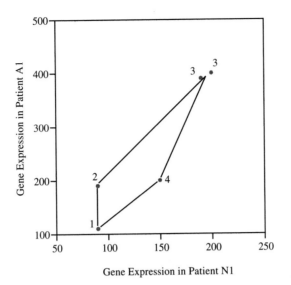

Fig. 5.4   SOM clustering of genes into a two-by-two grid, resulting in four clusters.

5.4   DISTANCE MEASURES

In addition to calculating the Euclidean distance, there are a number of other ways to calculate distance between two genes. When these are combined with different ways of normalizing your data, the choice of normalization and distance measure can become rather confusing. Here I will attempt to show how the different distance measures relate to each other and what effect, if any, normalization of the data has. Finally I will suggest a good choice of distance measure for expression data.

The Euclidean distance between two points a and b in N-dimensional space is defined as

$$\sqrt{\sum_{i=1}^{N}(a_i - b_i)^2}$$

where i is the index that loops over the dimensions of N, and the Σ sign indicates that the squared distances in each dimension should be summed before taking the square root of those sums. Figure 5.5 shows the Euclidean distance between two points in two-dimensional space.

Instead of calculating the Euclidean distance, you can also calculate the angle between the vectors that are formed between the data point of the gene and the center of the coordinate system. For gene expression, *vector angle* (Fig. 5.5) often performs better because the trend of a regulation response is

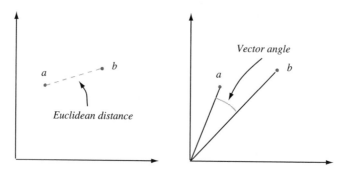

Fig. 5.5   Euclidean distance and vector angle between points a and b in two-dimensional space.

more important than its magnitude. Vector angle α between points a and b in N-dimensional space is calculated as:

$$\cos \alpha = \frac{\sum_{i=1}^{N} a_i b_i}{\sqrt{\sum_{i=1}^{N} a_i^2}\sqrt{\sum_{i=1}^{N} b_i^2}}$$

Finally, a widely used distance metric is the Pearson correlation coefficient:

$$\frac{\sum_{i=1}^{N}(a_i - \bar{a})(b_i - \bar{b})}{\sqrt{\sum_{i=1}^{N}(a_i - \bar{a})^2}\sqrt{\sum_{i=1}^{N}(b_i - \bar{b})^2}}$$

You can see that the only difference between vector angle and Pearson correlation is that the means ($\bar{a}$ and $\bar{b}$) have been subtracted before calculating the Pearson correlation. So taking the vector angle of a means-normalized data set (each gene has been centered around its mean expression value over all conditions) is the same as taking the Pearson correlation.

An example will illustrate this point. Let us consider two genes, a and b, that have the expression levels $a = (1, 2, 3, 4)$ and $b = (2, 4, 6, 8)$ in four experiments. They both show an increasing expression over the four experiments, but the magnitude of response differs. The Euclidean distance is 5.48, while the vector angle distance $(1 - \cos \alpha)$ is zero and the Pearson distance $(1 - PearsonCC)$ is zero. I would say that because the two genes show exactly the same trend in the four experiments, the vector angle and Pearson distance make more sense in a biological context than the Euclidean distance.

Table 5.2   Expression readings of four genes in six patients

Gene	Patient					
	N_1	N_2	A_1	A_2	B_1	B_2
a	90	110	190	210	290	310
b	190	210	390	410	590	610
c	90	110	110	90	120	80
d	200	100	400	90	600	200

Table 5.3   Euclidean distance matrix between four genes.

Gene	Gene			
	a	b	c	d
a	0.00	5.29	3.20	4.23
b	5.29	0.00	8.38	5.32
c	3.20	8.38	0.00	5.84
d	4.23	5.32	5.84	0.00

Table 5.4   Vector angle distance matrix between four genes.

Gene	Gene			
	a	b	c	d
a	0.00	0.02	0.42	0.52
b	0.02	0.00	0.41	0.50
c	0.42	0.41	0.00	0.51
d	0.52	0.50	0.51	0.00

5.4.1 Example: Comparison of Distance Measures

Let us try the different distance measures on our little example of four genes
from six patients (Table 5.2). We can calculate the Euclidean distances (see
Section 12.3 for details on how to do this) between the four genes. The
pairwise distances between all genes can be shown in a *distance matrix*
(Table 5.3) where the distance between gene a and a is zero, so the pairwise
identities form a diagonal of zeros through the matrix. The triangle above
the diagonal is a mirror image of the triangle below the diagonal because the
distance between genes a and b is the same as the distance between genes b
and a.

This distance matrix is best visualized by clustering as shown in Figure 5.6,
where it is compared with clustering based on vector angle distance (Table 5.4)
and a tree based on Pearson correlation distances (Table 5.5).

Table 5.5 Pearson distance matrix between four genes.

Gene	Gene			
	a	*b*	*c*	*d*
a	0.00	0.06	1.45	1.03
b	0.06	0.00	1.43	0.98
c	1.45	1.43	0.00	0.83
d	1.03	0.98	0.83	0.00

The clustering in Figure 5.6 is nothing but a two-dimensional visualization of the four-by-four distance matrix. What does it tell us? It tells us that vector angle distance is the best way to represent gene expression responses. Genes *a*, *b*, and *d* all have increasing expression over the three patient categories, only the magnitude of the response and the error between replicates differs. The vector angle clustering has captured this trend perfectly, grouping *a* and *b* close together and *d* nearby. Euclidean distance has completely missed this picture, focusing only on absolute expression values, and placed genes *a* and *b* furthest apart. Pearson correlation distance has done a pretty good job, capturing the close biological proximity of genes *a* and *b*, but it has normalized the data too heavily and placed gene *d* closest to gene *c*, which shows no trend in the disease at all.

It is also possible to cluster in the other dimension, clustering patients instead of genes. Instead of looking for genes which show a similar transcriptional response to the progression of a disease, we are looking for patients which have the same *profile* of expressed genes. If two patients have exactly the same stage of a disease we hope that this will be reflected in identical expression of a number of key genes. Thus, we are not interested in the genes that are not expressed in any patient, are unchanged between patients, or show a high error. So it makes sense to remove those genes before clustering

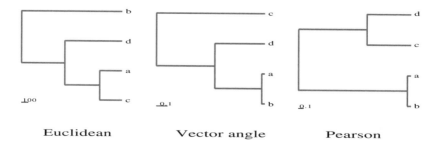

Euclidean   Vector angle   Pearson

Fig. 5.6 Hierarchical clustering of distances (with three different distance measures) between genes in the example.

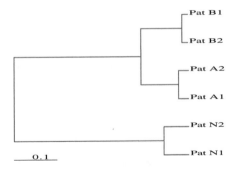

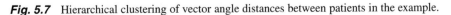

Fig. 5.7 Hierarchical clustering of vector angle distances between patients in the example.

patients. But that requires applying a t-test or ANOVA on the data and in order to do so you have to put the data into the categories you presume the patients fall into. That could obscure trends in the data that you have not yet considered. In practice, you could try clustering both on the full data and on data cleaned by t-test, ANOVA, or another method.

The data in our little example contains too much noise from gene c to cluster on the complete set of genes. If an ANOVA is run on the three patient categories (Section 3.5) that only leaves genes a and b, and the hierarchical vector angle distance clustering based on those two genes can be seen in Figure 5.7.

5.5 NORMALIZATION

The difference between vector angle distance and Pearson correlation comes down to a means normalization. There are two other common ways of normalizing the expression level of a gene—length normalization and S.D. normalization:

- Means: calculate mean and subtract from all numbers.

- Length: calculate length of gene vector and divide all numbers by that length.

- S.D.: calculate standard deviation and divide all numbers by it.

For each of these normalizations it is important to realize that it is performed on each gene in isolation; the information from other genes is not taken into account. Before you perform any of these normalizations, it is important that you answer this question: Why do you want to normalize the data in that way? Remember, you have already scaled the data, so expression readings should be comparable (scaling, by the way, is a form of global normalization). In general, normalization affects Euclidean distances to a large extent, it affects vector angles to a much smaller extent, and it hardly ever affects Pearson distances because the latter metric is normalized already. My suggestion for biological data is to use vector angle distance on non-normalized expression data for clustering.

5.6   VISUALIZATION OF CLUSTERS

Clusters are traditionally visualized with trees (Figures 5.7, 5.6, 5.2, and 5.1). Note that information is lost in going from a full distance matrix to a tree visualization of it. Different trees can represent the same distance matrix.

In DNA chip analysis it has also become common to visualize the gene vectors by representing the expression level or fold change in each experiment with a color-coded matrix. Figure 5.8 shows such a visualization of gene expression data using both a tree and a color matrix using the ClustArray software (Section 12.3).

5.6.1   Example: Visualization of Gene Clusters in Bladder Cancer

Figure 5.8 is a visualization of the most important genes (selected by their covariance to the progression of the disease) in DNA microarray measurements in bladder cancer patients.

5.7   SUMMARY

Cluster analysis groups genes according to how they behave in experiments. For gene expression, measuring similarity of gene expression using the vector angle between expression profiles of two genes makes most sense.

Normalization of your data matrix (of genes versus experiments) can be performed in either of two dimensions. If you normalize columns you normalize the total expression level of each experiment. A normalization of experiments to yield the same sum of all genes is referred to in this book as scaling and is described in Section 3.2. Such a normalization is essential before comparison of experiments, but a multifactor scaling with a spline or a polinomium is even better.

Normalization of genes in the other dimension may distort the scaling of experiments that you have performed (if you sum the expression of all genes in an experiment after a gene normalization, it will no longer add up to the same number). Also, normalization of genes before calculating vector angle is usually not necessary. Therefore, Pearson correlation is not quite as good a measure of similarity as vector angle.

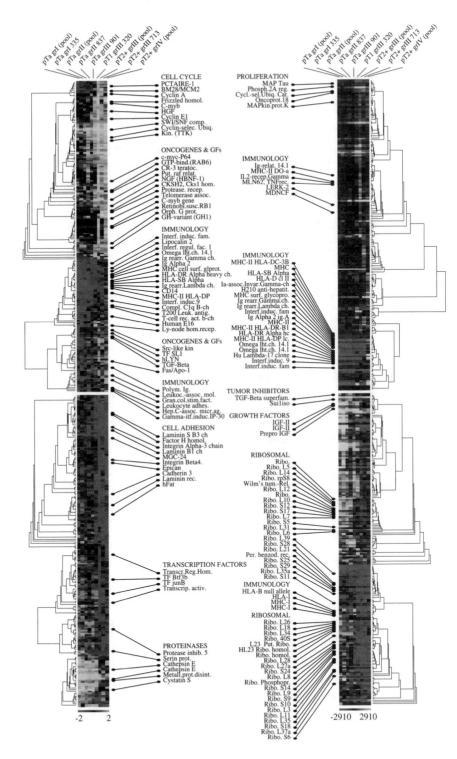

Fig. 5.8 Hierarchical clustering of genes (rows) expressed in bladder cancers (columns). Yellow fields show up-regulation of genes (absolute difference in right panel, logfold change in left panel), blue fields show down-regulation of genes. (Figure by Christopher Workman using ClustArray software on data from Thykjaer, et al., (2001) and postprocessing with Adobe Illustrator). (See color plate.)

5.8   FURTHER READING

Thykjaer, T., Workman, C., Kruhøffer, M., Demtröder, K., Wolf, H., Andersen, L. D., Frederiksen, C. M., Knudsen, S., and Ørntoft, T. F. (2001). Identification of gene expression patterns in superficial and invasive human bladder cancer. *Cancer Research* 61:2492–2499.

Clustering methods and cluster reliability

Getz, G., Levine, E., and Domany, E. (2000). Coupled two-way clustering analysis of gene microarray data. *Proc. Natl. Acad. Sci. USA* 97:12079–12084.

Hastie, T., Tibshirani, R., Eisen, M. B., Alizadeh, A., Levy, R., Staudt, L., Chan, W. C., Botstein, D., and Brown, P. (2000). Gene shaving as a method for identifying distinct sets of genes with similar expression patterns. *Genome Biol.* 1:RESEARCH0003.1–21.

Self-organizing tree algorithm: Herrero, J., Valencia, A., and Dopazo, J. (2001). A hierarchical unsupervised growing neural network for clustering gene expression patterns. *Bioinformatics* 17:126–136.

Kerr, M. K., and Churchill, G. A. (2001). Bootstrapping cluster analysis: Assessing the reliability of conclusions from microarray experiments. *Proc. Natl. Acad. Sci. USA* 98:8961–5.

Kohonen, T. (1995). *Self-Organizing Maps*. Berlin: Springer.

Clustering of time-series data: Michaels, G. S., Carr, D. B., Askenazi, M., Fuhrman, S., Wen, X., and Somogyi, R. (1998). Cluster analysis and data visualization of large-scale gene expression data. *Pacific Symposium on Biocomputing* 3:42–53.[1]

Sasik, R., Hwa, T., Iranfar, N., and Loomis, W. F. (2001). Percolation clustering: A novel algorithm applied to the clustering of gene expression patterns in dictyostelium development. *Pacific Symposium on Biocomputing* 6:335–347.[2]

Tamayo, P., Slonim, D., Mesirov, J., Zhu, Q., Kitareewan, S., Dmitrovsky, E., Lander, E. S., and Golub, T. R. (1999). Interpreting patterns of gene expression with self-organizing maps: Methods and application to hematopoietic differentiation. *Proc. Natl. Acad. Sci. USA* 96:2907–2912.

---

[1] Available online at http://psb.stanford.edu
[2] Available online at http://psb.stanford.edu

Tibshirani, R., Walther, G., Botstein, D., and Brown, P. (2000). Cluster validation by prediction strength. Technical report. Statistics Department, Stanford University.[3]

Xing, E. P., and Karp, R. M. (2001). CLIFF: Clustering of high-dimensional microarray data via iterative feature filtering using normalized cuts. *Bioinformatics* 17(Suppl 1):S306–S315.

Yeung, K. Y., Haynor, D. R., and Ruzzo, W. L. (2001). Validating clustering for gene expression data. *Bioinformatics* 17:309–18.

Yeung, K. Y., Fraley, C., Murua, A., Raftery, A. E., and Ruzzo, W. L. (2001) Model-based clustering and data transformations for gene expression data. *Bioinformatics* 17:977–87.

6

Beyond Cluster Analysis

6.1 FUNCTION PREDICTION

Genes that appear in the same cluster have similar transcription response to different conditions. It is likely that this is caused by some commonality in function or role. If a cluster is populated by genes with known function—and that function is similar—you can infer the function of orphan genes in the same cluster.

Another valuable tool to assigning function, in particular for clusters where there are no genes with known function, is function prediction. Function prediction enters the scene where there is no sequence homology to proteins with known function. Instead, a number of properties and predicted features of the protein can be used to predict a likely function class (Jensen, et al., 2002). It turns out that proteins with similar function also share some similarities in amino acid sequence length, posttranslational modification, cellular destination signal, and so on. Taken separately, each of these features is a weak predictor of function category. Taken together, a sufficiently large number of features can be used to make fairly accurate predictions of function class.

6.2   DISCOVERY OF REGULATORY ELEMENTS IN PROMOTER REGIONS

If a number of genes share a regulatory response to a number of stimuli it is reasonable to assume that they do so because they share a binding site for a transcription factor in their promoter.

The ClustArray web-based clustering software (Section 12.3) allows you to select the genes in a cluster and search their upstream promoter region for such common regulatory elements that can account for the similarity in transcription response. This works only for organisms where the promoter regions of the genes on the chip are known and included in a database. Currently, databases of promoter regions in human and *Saccharomyces cerevisiae* are available at the website.

Even when you have the promoter sequence, finding common regulatory elements is inherently complicated because of the degeneracy of such elements. You can use software such as *saco-patterns* (Jensen and Knudsen, 2000) to search for patterns that are fully conserved. An example of a fully conserved pattern is AGCTTAGG. Such a search is reasonably fast, simple, and deterministic, because it is possible to search for all possible patterns up to a given length. But it will miss all those patterns that are not conserved enough to be picked up by a single pattern such as AGCTTAGG. Transcription factor binding sites are typically *degenerate*; they tolerate some variation in sequence at some locations in the site. The problem is that there is an infinite number of possible degenerate sites. Still, software solutions have been developed for this problem. Degenerate patterns can be searched with software like *ann-spec* (Workman and Stormo, 2000), but it is sensitive to the choice of parameters, and it will not give the same result every time you run it. It uses a probabilistic Gibbs sampling approach to guess parameters for a weight matrix that describes the regulatory elements.

Lawrence's Gibbs sampler (Neuwald and Lawrence, 1995; Lawrence, et al., 1993), uses a similar strategy.

Running any of these methods to discover regulatory elements will lead you into an assessment of the significance of any discoveries. There are two good ways of assessing this. First, you can look for the occurrence of the discovered element in a background set, either in a set of promoters known not to contain the element, or in a set of all promoters in that organism, where you can assume that most promoters do not contain the element. Then you can compare the frequency of elements in the promoters in your positive set to the frequency of promoters in the background set and perform a statistical analysis (sampling without replacement) to calculate the probability that both sets have the same occurrence of the elements. If they do, then you have not found a biologically relevant element. Remember to correct for multiple testing (Bonferroni) before evaluating probabilities (see Section 3.5). Saco-patterns includes a statistical evaluation with Bonferroni correction.

Confirmation of a biologically relevant signal may also come from an observation that those genes in the background set that do contain the element actually are additional members of the pathway or function class of the positive set.

Finally, a way of assessing significance is to plot a histogram of positional preference of the signal relative to the transcription start site. Any obvious preference is a confirmation of biological significance, while an absence does not rule out significance.

If you perform any of the tests described above, be sure to perform them on promoter extracts of identical length to avoid artifacts of analysis.

6.2.1   Example 1: Discovery of Proteasomal Element

If you take all 6269 ORFs annotated in the GenBank file of *Saccharomyces cerevisiae* and extract 200 bp starting 300 bp upstream of the ORF, you cover most promoter regions in the organism pretty well. If you divide these 6269 promoter regions into those that have been annotated as related to the proteasome (31) and those that have not (6238), you have a positive set and a background set, respectively. If you run saco-patterns using these sets as positive and background, it finds the sequence GGTGGCAAA present in 25 of the positive set and 26 of the background set. That is such a vast overrepresentation that the probablity that it is not significant—even after correction for multiple testing—is less than 10^{-10}. Of the 26 apparent false positives in the background set two are proteases and three are genes with relation to ubiquitin, all of which could very well be coregulated with proteasomes (Jensen and Knudsen, 2000).

Note that in this example we did not use expression analysis such as *t*-test or clustering to generate a positive set of promoters. We used functional annotation. You can use any method to generate a positive set and then search for patterns overrepresented in that set.

6.2.2   Example 2: Rediscovery of Mlu Cell Cycle Box (MCB)

Using the yeast promoter regions from the previous example, but instead sorting them by the expression in one of the cell cycle experiments (Spellman, et al., 1998), allows identification of patterns that are correlated with expression: instead of dividing promoter regions into a positive and negative set, we look for patterns that are more frequent in up-regulated genes than in nonregulated genes or in down-regulated genes, or vice versa. There is a statistical test for this and it is called the Kolmogorov-Smirnov rank test (Jensen and Knudsen, 2000). Figure 6.1 shows the distribution of genes that contain such a pattern, the Mlu cell cycle Box (MCB). The well-known MCB pattern, ACGCGT, was discovered to be significant by saco-patterns testing all possible patterns up to length 8 in the cell cycle experiment.

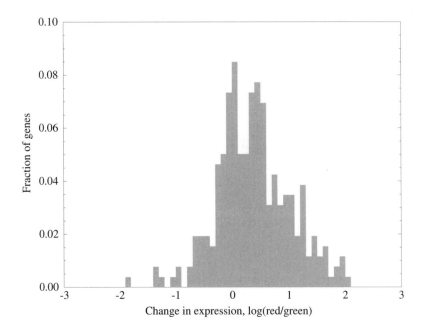

Fig. 6.1   Logfold distribution of all yeast genes (open bars) in a cell cycle experiment. Logfold distribution of genes containing Mlu cell cycle box (shaded bars) in the same experiment. Drawing by Lars Juhl Jensen based on data from Jensen and Knudsen (2000).

6.3   INTEGRATION OF DATA

The analysis of data from an expression experiment becomes much more powerful if information about the function of genes and knowledge of the promoter elements controlling expression of the genes are included.

Such an integration of data can be done manually on a small scale. If you have a small group of genes with significant difference in expression between two conditions or you have a cluster of genes behaving the same way in a number of experiments, you can subject them to further analysis. The obvious first choice is to look at the functional annotation of the genes to discover clues of a pathway they may all be a part of. That pathway may not become evident until you actually read the Medline abstract(s) for the gene(s) in question. Second, you can run a promoter analysis as described in Section 6.2. If genes that show common regulation also participate in a common pathway or function, and have a regulatory element in common in their promoter, you have a pretty strong case—in more sense than one, because combining results from different analyses may increase the statistical significance of the finding.

The problem is that such an analysis takes a long time, and you can only do it manually for a small number of genes. What is really needed is a computer

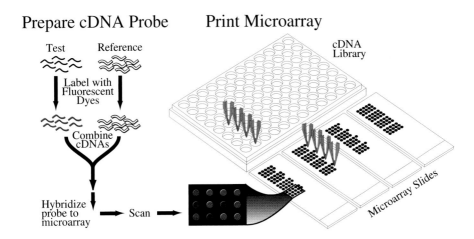

Fig. 1.4   The spotted array technology. A robot is used to transfer probes in solution from a microtiter plate to a glass slide where they are dried. Extracted mRNA from cells is converted to cDNA and labeled fluorescently. Sample is labeled red and control is labeled green. After mixing, they are hybridized to the probes on the glass slide. After washing away unhybridized material, the chip is scanned with a confocal laser and the image analyzed by computer.

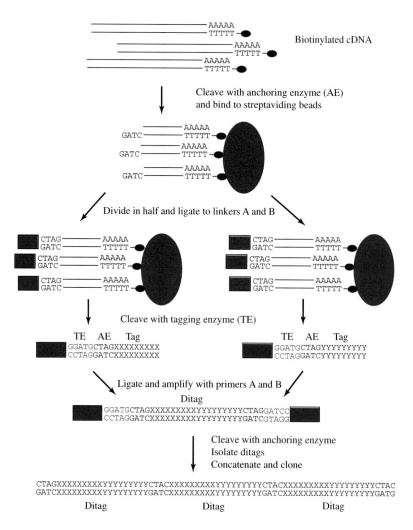

Fig. 1.5 Schematic overview of SAGE methods (based on Velculescu, et al. 1995).

Fig. 1.6 cDNA microarray of genes affected by HIV infection.

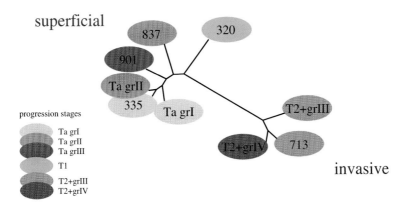

Fig. 5.2 Hierarchical clustering of bladder cancer patients. The clustering was based on expression measurements from a DNA array hybridized with mRNA extracted from a biopsy. (Christopher Workman, based on data published in Thykjaer, et al. (2001)).

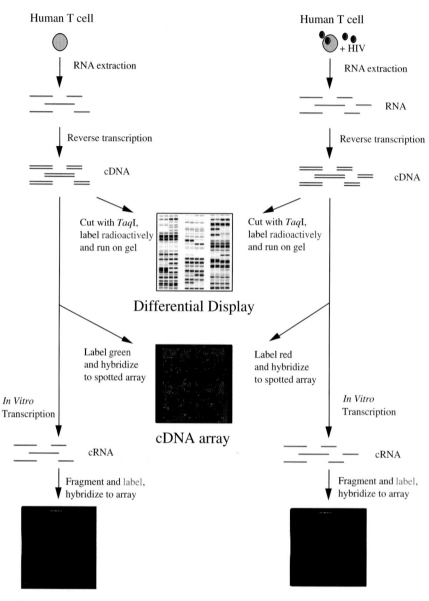

Fig. 1.8 Overview of methods for comparing mRNA populations in cells from two different conditions.

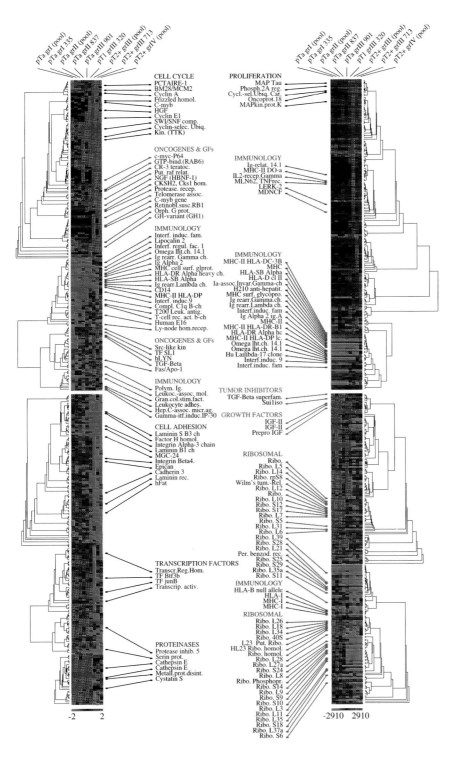

Fig. 5.8 Hierarchical clustering of genes (rows) expressed in bladder cancers (columns). Yellow fields show up-regulation of genes (absolute difference in right panel, logfold change in left panel), blue fields show down-regulation of genes. (Figure by Christopher Workman using ClustArray software on data from Thykjaer, et al., (2001) and postprocessing with Adobe Illustrator).

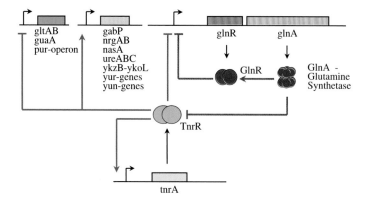

Fig. 7.2 Known regulatory network in *Bacillus subtilis*. Each line ending in a bar represents a deduced negative regulatory effect. Each line ending in an arrow represents a deduced positive regulatory effect. (Hanne Jarmer and Carsten Friis.)

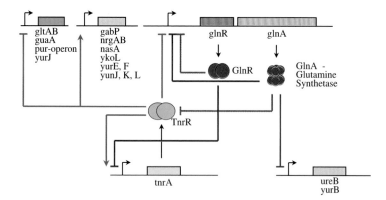

Fig. 7.3 Regulatory network reverse engineered from real steady-state data. Each line ending in a bar represents a deduced negative regulatory effect. Each line ending in an arrow represents a deduced positive regulatory effect. (Hanne Jarmer and Carsten Friis.)

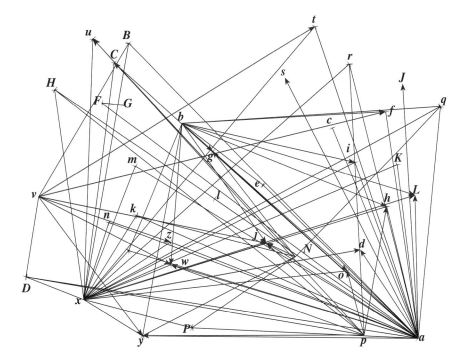

Fig. 7.4 Regulatory network reverse engineered from real steady-state data. Each letter represents a gene or a cluster of genes. Each line ending in a bar represents a negative regulatory effect. Each line ending in an arrow represents a positive regulatory effect. (Output from computer software developed by Carsten Friis.)

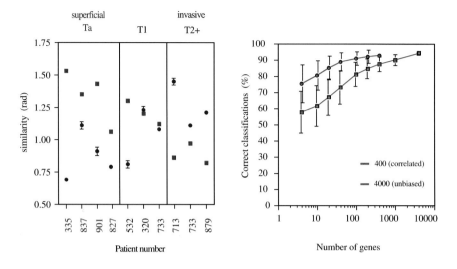

Fig. 8.1 Classifier of bladder cancers based on expression array. Left: Vector angle between patient and reference pool. Each three-digit number on the bottom refers to a patient. The angle between that patient and the two reference pools (squares, Ta pool; circles T2 pool) is indicated. The angle is always smallest (the similarity is greatest) to the pool with the same type of cancer. The intermediate type, T1, for which there is no reference pool, is sometimes more similar to one reference, sometimes more similar to another reference. Error bars have been added to show variation due to choice of reference pool, of which several were available. Right: the performance of the classifier as a function of the number of genes used for classification. Top curve: genes chosen among those 400 genes maximally covarying with the disease. Bottom curve: genes chosen at random from all 4000 genes detected as present in at least one patient.

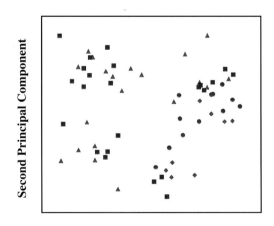

First Principal Component

Fig. 8.2 Principal component analysis of 63 small, round blue cell tumors. Different symbols are used for each of the four categories as determined by classical diagnostics tests.

algorithm that looks for these correlations between transcriptome, proteome, and genome in a systematic way and reports all significant correlations. One approach is to cluster not only on expression data but on functional annotation and promoter annotation as well. But how do you annotate a promoter and how do you cluster on a functional annotation? That is where there is still some work left to be done (see Masys, et al., 2001; and Jenssen, et al., 2001).

6.4 SUMMARY

It is beyond the cluster analysis that the real data mining takes place: you can mine your data for promoter elements involved in the regulation that you observe, you can mine for novel functions of orphan proteins, you can mine for novel regulatory relationships between genes under study. In the future, these analyses should be combined to increase their power in detecting subtle relationships that may today be obscured by noise in your data.

6.5 FURTHER READING

Jensen, L. J., Gupta, R., Blom, N., Devos, D., Tamames, J., Kesmir, C., Nielsen, H., Stærfeldt, H. H., Rapacki, K., Workman, C., Andersen, C. A. F., Knudsen, S., Krogh, A., Valencia, A., and Brunak., S. (2002). Ab initio prediction of human orphan protein function from post-translational modifications and localization features. Submitted.

Promoter element discovery tools

Brazma, A., Jonassen, I., Vilo, J., and Ukkonen, E. (1998). Predicting gene regulatory elements in silico on a genomic scale. *Genome Research* 8:1202–1215.

Bussemaker, H. J., Li, H., and Siggia, E. D. (2000). Building a dictionary for genomes: Identification of presumptive regulatory sites by statistical analysis. *Proc. Natl. Acad. Sci. USA* 97:10096–10100.

Birnbaum, K., Benfey, P. N., and Shasha, D. E. (2001). Cis element/transcription factor analysis (cis/TF): A method for discovering transcription factor/cis element relationships. *Genome Res.* 11:1567–1573.

Chiang, D. Y., Brown, P. O., and Eisen, M. B. (2001). Visualizing associations between genome sequences and gene expression data using genome-mean expression profiles. *Bioinformatics* 17(Suppl 1):S49–S55.

Claverie, J.-M. (1999). Computational methods for the identification of differential and coordinated gene expression. *Hum. Mol. Genet.* 8:1821–1832.

Fujibuchi, W., Anderson, J. S. J., and Landsman, D. (2001). PROSPECT improves cis-acting regulatory element prediction by integrating expression profile data with consensus pattern searches. *Nucleic Acids Res.* 29:3988–3996.

Jensen, L. J., and Knudsen, S. (2000). Automatic discovery of regulatory patterns in promoter regions based on whole cell expression data and functional annotation. *Bioinformatics* 16:326–333.

Lawrence, C. E., Altschul, S. F., Boguski, M. S., Liu, J. S., Neuwald, A. F., and Wootton, J. C. (1993). Detecting subtle sequence signals: A Gibbs sampling strategy for multiple alignment. *Science* 262:208–214.

Liu, X., Brutlag, D. L., and Liu, J. S. (2001). BioProspector: Discovering conserved DNA motifs in upstream regulatory regions of co-expressed genes. *Pacific Symposium on Biocomputing* 6:127–138.[1]

Neuwald, A. F., Liu, J. S., and Lawrence, C. E. (1995). Gibbs motif sampling: Detection of bacterial outer membrane protein repeats. *Protein Science* 4:1618–1632.

Spellman, P., Sherlock, G., Zhang, M., Lyer, V., Anders, K., Eisen, M., Brown, P., Botstein, D., and Futcher, B. (1998). Comprehensive identification of cell cycle-regulated genes of yeast *S. cerevisiae* by microarray hybridization. *Mol. Biol. Cell* 9:3273–3297.

Wolfsberg, T. G., Gabrielian, A. E., Campbell, M. J., Cho, R. J., Spouge, J. L., and Landsman, D. (1999). Candidate regulatory sequence elements for cell cycle-dependent transcription in *Saccharomyces cerevisiae*. *Genome Res.* 9:775–792.

Workman, C., and Stormo, G.D. (2000) ANN-Spec: A method for discovering transcription factor binding sites with improved specificity. *Pacific Symposium on Biocomputing 2000*.[2]

Integration of data

Ideker, T., Thorsson, V., Ranish, J. A., Christmas, R., Buhler, J., Eng, J. K., Bumgarner, R., Goodlett, D. R., Aebersold, R., and Hood, L. (2001). Integrated genomic and proteomic analyses of a systematically perturbed metabolic network. *Science* 292:929–34.

Jenssen, T. K., Laegreid, A., Komorowski, J., and Hovig, E. (2001). A literature network of human genes for high-throughput analysis of gene expression. *Nat Genet.* 28:21–8.

---

[1] Available online at http://psb.stanford.edu
[2] Available online at http://psb.stanford.edu

Masys, D. R., Welsh, J. B., Lynn Fink, J., Gribskov, M., Klacansky, I., and Corbeil, J. (2001). Use of keyword hierarchies to interpret gene expression patterns. *Bioinformatics*. 17:319–26.[3]

Noordewier, M. O., and Warren, P. V. (2001). Gene expression microarrays and the integration of biological knowledge. *Trends. Biotechnol.* 19:412–5.

Tanabe, L., Scherf, U., Smith, L. H., Lee, J. K., Hunter, L., and Weinstein, J. N. (1999). MedMiner: An internet text-mining tool for biomedical information, with application to gene expression profiling. *BioTechniques* 27:1210–1217.[4]

Rain, J. C., Selig, L., De Reuse, H., Battaglia, V., Reverdy, C., Simon, S., Lenzen, G., Petel, F., Wojcik, J., Schachter, V., Chemama, Y., Labigne, A., and Legrain, P. (2001). The protein-protein interaction map of *Helicobacter pylori*. *Nature* 409:211–5.

Zhu, J., and Zhang, M. Q. (2000). Cluster, function and promoter: Analysis of yeast expression array. *Pacific Symposium on Biocomputing* 5:476–487.[5]

---

[3] Web-based software available at http://array.ucsd.edu/hapi/
[4] Web version available at http://discover.nci.nih.gov/textmining/filters.html
[5] Available online at http://psb.stanford.edu

7

Reverse Engineering of Regulatory Networks

One gene can affect the expression of another gene by binding of the gene product of one gene to the promoter region of another gene. Looking at more than two genes, we refer to the *regulatory network* as the regulatory interactions between the genes.

If we have a large number of measurements of the expression level of a number of genes, we should be able to model or reverse engineer the regulatory network that controls their expression level. The problem can be attacked in two fundamentally different ways: using time-series data and using steady-state data of gene knockouts.

7.1 THE TIME-SERIES APPROACH

The expression level of a certain gene at a certain time point can be modeled as some function of the expression levels of all other genes at all previous time points.

The problem is that you usually have many more genes than you have time points! That means that you have a dimensionality problem: there are too many parameters and too few equations to estimate them. If you have g genes, there are g^2 possible connections between them and you would need at least g^2 linearly independent equations to determine all of them.

Different solutions to this problem have been developed. One is the linear modeling approach by van Someren et al. (2000). They reduce the dimensionality of the problem by first removing all genes that do not show a significant change in expression through the experiment. Then they cluster the genes to group those that behave the same way. There is no reason mathematically to distinguish between two genes if you cannot distinguish their transcription response. Then they try to build a linear model of the remaining gene clusters. The basic linear model follows the assumption that the activity of a gene x equals the weighted sum of the activities of all N genes at the previous time point $(t - 1)$:

$$x_j(t) = \sum_{i=1}^{N} r_{i,j} x_i(t - 1)$$

where $r_{i,j}$ is a weight factor representing how gene i affects gene j, positively or negatively. Preliminary data indicate that they still need a few more experimental data points to solve the models exactly for the yeast cell cycle experiments than those that were available at the time.

Another solution to the dimensionality problem was developed by Holter et al. (2000), who used singular value decomposition (similar to principal component analysis, see Section 4.1) to reduce the dimensionality before solving the interaction matrix. That leaves fewer independent genes and makes it easier to find their interactions.

An entirely different approach is Bayesian networks, where the problem is simplified to one of genes that are up-regulated and genes that are down-regulated (Friedman, et al., 2000). That still leaves a dimensionality problem, but they try to estimate probabilistic networks that fit the data and look for results that are common in different models that fit the same data. They have shown some success in extracting central regulatory pathways in yeast.

7.2 THE STEADY-STATE APPROACH

A particularly attractive approach is the steady-state model, where the effect of deleting a gene on the expression of other genes is measured. If the expression of gene b increases after deletion of gene a, it can be inferred that gene a repressed, either directly or indirectly, the expression of gene b. If the expression of gene b decreases after deletion of gene a, it can be inferred that gene a enhanced, either directly or indirectly, the expression of gene b. With a large DNA microarray, it is possible to determine all consequences of deletion of gene a.

Such results will give valuable information about the regulatory network that the deleted gene is involved in (Ideker, et al., 2000). As compendiums of expression profiles of gene deletions become available (Hughes, et al., 2000) the steady-state model is a very promising tool for extracting regulatory networks.

7.3 LIMITATIONS OF NETWORK MODELING

It must never be forgotten, however, that the genetic network approaches so far all ignore those regulatory interactions that take place at the protein-protein level. A lot of cellular regulation, for example of the cell cycle, takes place through phosphorylation and dephosphorylation of proteins. In the future, regulatory network models must include such information, for example, by inclusion of protein-protein interaction maps (Rain, et al., 2001) determined by using the yeast two-hybrid assay. What is also needed is a way to combine prior biological knowledge of regulatory networks (Tanay and Shamir, 2001), information deduced from time-series experiments, and information deduced from steady-state experiments. If information from each can be represented as a matrix of interactions between genes, then the three matrices can be summed and the regulatory network deduced from that. The disadvantage of running the three methods independently, however, is that the solid information from prior knowledge and direct deletion in steady-state experiments can be useful in determining which time series models best fit data from all three domains. Optimally, the information from prior knowledge and steady state models should be used when deriving time-series models.

The regulatory network can be visualized by drawing a box for each gene which has interactions above a cutoff threshold with other genes. Next all the interactions (above threshold) are drawn as lines between the boxes (line width can be scaled by interaction strength; postitive and negative interactions can be distinguished by lines ending in an arrow and in a bar, respectively). For more than 100 interactions or so, this visualization quickly becomes unwieldy, and subnetworks have to be extracted and drawn.

7.4 EXAMPLE 1: STEADY-STATE MODEL

Let us take an example of four genes, a, b, c, and d. When we delete gene a, we find that the expression of gene b and d decreases. We conclude that gene

Table 7.1 Interaction matrix between four genes.

Gene	Gene a	b	c	d
a		$+$		$+$
b				$+$
c	$-$	$-$		$-$
d				

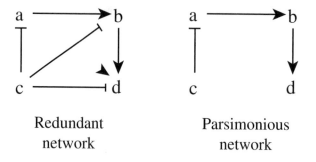

Redundant
network

Parsimonious
network

Fig. 7.1 Regulatory networks deduced from experimentally produced interaction matrix. Arrows mean positive regulation; bars mean negative regulation.

a has a stimulatory effect, directly or indirectly, on genes b and d. We can represent this information in an interaction matrix (Table 7.1) Note that there is a direction to each interaction. Rows represent genes which are deleted; columns represent genes whose expression is changed as a result. In the matrix we have included the results from two other deletion experiments: gene b, which led to a decrease in d, and gene c, which led to an increase in a, b, and d.

From this matrix we can now draw up a redundant genetic network which represents all the interactions between genes as arrows (positive regulation) or bars (negative regulation) (Figure 7.1).

This regulatory network is redundant in that it contains both direct and indirect regulations. There are several paths between two genes. What we now wish to deduce is the parsimonious network—the smallest and simplest network that is able to explain the experimental observations. If there is more than one path between two genes, we want to delete those that are not necessary to explain the results. That is achieved by eliminating for each pair of genes all but the longest path (involving most genes) if that path is still able to explain the regulatory effect observed.

Between genes c and b there are two possible paths that both have the same effect, so we remove the shortest path, the direct path between c and b. Between genes a and d there are two paths that both have the same effect, so we remove the shortest one, the direct path between a and d. Finally, between genes c and d, we remove the direct path between c and d. The resulting network is the simplest network that explains the data.

7.5   EXAMPLE 2: STEADY-STATE MODEL ON REAL DATA

The approach decribed above was applied to knockout mutants in the regulation of nitrogen metabolism in *Bacillus subtilis* (Figure 7.2, Jarmer, et. al.,

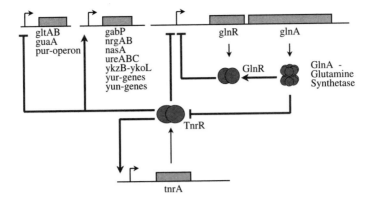

Fig. 7.2 Known regulatory network in *Bacillus subtilis*. Each line ending in a bar represents a negative regulatory effect. Each line ending in an arrow represents a positive regulatory effect. (Hanne Jarmer and Carsten Friis. See color plate.)

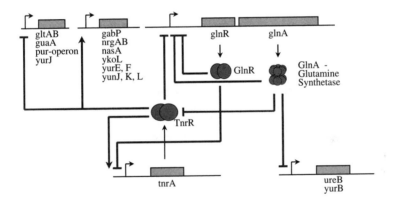

Fig. 7.3 Regulatory network reverse engineered from real steady-state data. Each line ending in a bar represents a deduced negative regulatory effect. Each line ending in an arrow represents a deduced positive regulatory effect. (Hanne Jarmer and Carsten Friis. See color plate.)

(2002)). Expression data was filtered for significance through a *t*-test. Genes were clustered into groups that show the same response in all experiments. From the interaction matrix a redundant network was generated and reduced to a parsimonious network. Figure 7.3 shows the resulting network as output by a computer program.

When compared to the known biological system shown in Figure 7.2, it is evident that the computer only missed the protein-protein interactions. But the computer discovered novel gene regulations in this system.

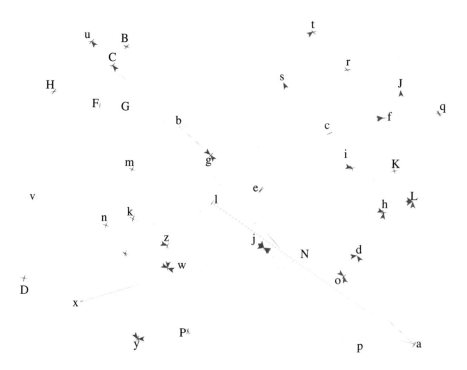

Fig. 7.4   Regulatory network reverse engineered from real steady-state data. Each letter represents a gene or a cluster of genes. Each line ending in a bar represents a negative regulatory effect. Each line ending in an arrow represents a positive regulatory effect. (Output from computer software developed by Carsten Friis. See color plate.)

7.6   EXAMPLE 3: STEADY-STATE MODEL ON REAL DATA

The approach decribed above was applied to another real expression dataset filtered for significance through an ANOVA. Genes were clustered into groups that show the same response in all experiments. From the interaction matrix a redundant network was generated and reduced to a parsimonious network. Figure 7.4 shows the resulting network as output by a computer program.

7.7   EXAMPLE 4: LINEAR TIME-SERIES MODEL

Let us try to deduce a network from time-series data as well. This is a little more complicated and involves some matrix algebra. Suppose we have conducted the experiment shown in Figure 7.5. At time zero we induce gene c with a substrate or other induction of its promoter. At times 1, 2, 3, and 4 we follow the expression level of gene c and three other genes, a, b, and d,

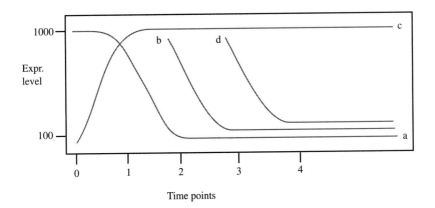

Fig. 7.5 Time-series experiment with four genes.

and see how they change in response to the induction of gene c. We represent the expression of each gene at each time point as the logarithm (base 10) of the fold change relative to time zero. So gene a at time 0 is expressed $\log_{10}(1000/1000) = 0$, at time 1 is expressed $\log_{10}(1000/1000) = 0$, at time 2, 3, and 4 is expressed $\log_{10}(1000/100) = -1$. Gene c has expression levels 0, 1, 1, 1, 1 at time points 0, 1, 2, 3, and 4. We can put these logfold change numbers in an expression matrix (Table 7.2). To solve the regulatory network we plug into the formula mentioned above:

$$x_j(t) = \sum_{i=1}^{N} r_{i,j} x_i(t-1)$$

where $j = a, b, c, d$, and $i = a, b, c, d$, and $t = 1, 2, 3, 4$. This system of linear equations can be solved with standard techniques such as Gaussian elimination or singular value decomposition.

But we can illustrate the method by solving just the system of equations governing the regulation of gene a. At time $t = 4$, we have:

$$a(t = 4) = r_{b,a}b(t = 3) + r_{c,a}c(t = 3) + r_{d,a}d(t = 3)$$

Inserting the logfold values from our experiment time points 4 and 3, we get:

$$-1(t = 4) = r_{b,a} - 1 + r_{c,a}1 + r_{d,a}0$$

Likewise, we get for time points 3, 2, and 1:

$$-1(t = 3) = r_{b,a}0 + r_{c,a}1 + r_{d,a}0$$

$$-1(t = 2) = r_{b,a}0 + r_{c,a}1 + r_{d,a}0$$

$$0(t = 1) = r_{b,a}0 + r_{c,a}0 + r_{d,a}0$$

Here we have four equations (two are identical) with three unknowns, and they can be solved with standard methods. It turns out that there is only one solution, $r_{c,a} = -1$ and $r_{b,a} = r_{d,a} = 0$. You can test this by inserting the solution into the four equations. So we have deduced from this set of equations that gene a is negatively regulated by gene c.

Likewise, we find that gene b is positively regulated by gene a, and gene d is positively regulated by gene b. We can summarize these findings in a directional interaction matrix (Table 7.3). The interactions are nonredundant, so it is easy to draw up the simplest network that satisfies the interaction matrix (Figure 7.6). This network is identical to the one deduced in example 7.4.

It is rare, however, that the data are as well-behaved as in this hypothetical example. First, the time points must be sufficient to resolve unambiguously the order of events. So the separation between time points must be smaller

Table 7.2   Expression matrix for four genes.

			Time		
Gene	0	1	2	3	4
a	0	0	-1	-1	-1
b	0	0	0	-1	-1
c	0	1	1	1	1
d	0	0	0	0	-1

Table 7.3   Directional interaction matrix between four genes.

		Gene		
Gene	a	b	c	d
a		+		
b				+
c	−			
d				

than the time it takes to reach steady-state after induction of a gene. Second, the number of time points must be at least as large as the number of interactions between the genes studied.

Fig. 7.6 Regulatory network deduced from time-series interaction matrix. Arrows mean positive regulation; bars mean negative regulation.

7.8 FURTHER READING

Regulatory networks

Akutsu, T., Miyano, S., and Kuhara, S. (1999). Identification of genetic networks from a small number of gene expression patterns under the boolean network model. *Pacific Symposium on Biocomputing* 4:17–28.[1]

Chen, T., He, H. L., and Church, G. M. (1999). Modeling gene expression with differential equations *Pacific Symposium on Biocomputing* 4:29–40.[2]

D'haeseleer, P., Wen, X., Fuhrman, S., and Somogyi, R. (1999). Linear modeling of mRNA expression levels during CNS development and injury *Pacific Symposium on Biocomputing* 4:41–52.[3]

Friedman, N., Linial, M., Nachman, I., and Pe'er, D. (2000). Using Bayesian networks to analyze expression data. *Proc. Fourth Annual International Conference on Computational Molecular Biology (RECOMB)* 2000.

Hartemink, A. J., Gifford, D. K., Jaakkola, T. S., and Young, R. A. (2001). Using graphical models and genomic expression data to statistically validate models of genetic regulatory networks. *Pacific Symposium on Biocomputing* 6:422–433.[4]

---

[1] Available online at http://psb.stanford.edu
[2] Available online at http://psb.stanford.edu
[3] Available online at http://psb.stanford.edu
[4] Available online at http://psb.stanford.edu

Holter, N. S., Maritan, A., Cieplak, M., Fedoroff, N. V., and Banavar, J. R. (2000). Dynamic modeling of gene expression data. *Proc. Natl. Acad. Sci. USA* 98:1693–1698.

Hughes, T. R., Marton, M. J., Jones, A. R., Roberts, C. J., and Stoughton, R., et. al. (2000). Functional discovery via a compendium of expression profiles. *Cell* 102:109–26.

Ideker, T. E., Thorsson, V., and Karp, R. M. (2000). Discovery of regulatory interactions through perturbation: Inference and experimental design. *Pacific Symposium on Biocomputing* 5:305–16.[5]

Jarmer, H., Friis, C., Saxild, H. H., Berka, R., Brunak, S., and Knudsen, S. (2002). Inferring parsimonious regulatory networks in *B. subtilis*. *Pacific Symposium on Biocomputing* 2002. Poster presentation.

Kim, S., Dougherty, E. R., Bittner, M. L., Chen, Y., Sivakumar, K., Meltzer, P., and Trent, J. M. (2000) General nonlinear framework for the analysis of gene interaction via multivariate expression arrays. *J. Biomed. Opt.* 5:411–24.

Kim, S., Dougherty, E. R., Chen, Y., Sivakumar, K., Meltzer, P., Trent, J. M., and Bittner, M. (2000) Multivariate measurement of gene expression relationships. *Genomics* 15:201–9.

Liang, S., Fuhrman S., and Somogyi, R. (1998). REVEAL, A general reverse engineering algorithm for inference of genetic network architectures. *Pacific Symposium on Biocomputing* 3:18–29.[6]

Combination of temporal and steady-state data: Maki, Y., Tominaga, D., Okamoto, M., Watanabe, S., and Eguchi, Y. (2001). Development of a system for the inference of large scale genetic networks. *Pacific Symposium on Biocomputing* 6:446–458.[7]

Pe'er, D., Regev, A., Elidan, G., and Friedman, N. (2001). Inferring subnetworks from perturbed expression profiles. *Bioinformatics* 17 (Suppl. 1):S215–S224.

Samsonova, M. G., and Serov, V. N. (1999). NetWork: An interactive interface to the tools for analysis of genetic network structure and dynamics. *Pacific Symposium on Biocomputing* 4:102–111.[8]

---

[5] Available online at http://psb.stanford.edu
[6] Available online at http://psb.stanford.edu
[7] Available online at http://psb.stanford.edu
[8] Available online at http://psb.stanford.edu

van Someren, E. P., Wessels, L. F. A., and Reinders, M. J. T. (2000). Linear modeling of genetic networks from experimental data. *Proc. ISMB* 2000:355–366.

Segal, E., Taskar, B., Gasch, A., Friedman, N., and Koller, D. (2001). Rich probabilistic models for gene expression. *Bioinformatics* 17(Suppl 1):S243–S252.

Review of limitations and problems: Szallasi, Z. (1999). Genetic network analysis in light of massively parallel biological data acquisition. *Pacific Symposium on Biocomputing* 4:5–16.[9]

Roberts, C. J., Nelson, B., Marton, M. J., Stoughton, R., Meyer, M. R., Bennett, H. A., He, Y. D. D. , Dai, H. Y., Walker, W. L., Hughes, T. R., Tyers, M., Boone, C., and Friend, S. H. (2000). Signaling and circuitry of multiple MAPK pathways revealed by a matrix of global gene expression profiles. *Science* 287:873–880.

Tanay, A., and Shamir, R. (2001). Expansion on existing biological knowledge of the network: Computational expansion of genetic networks. *Bioinformatics* 17(Suppl 1):S270–S278.

Thieffry, D., and Thomas, R. (1998). Qualitative analysis of gene networks. *Pacific Symposium on Biocomputing* 3:77–88.[10]

Wahde, M., and Hertz, J. (2000). Coarse-grained reverse engineering of genetic regulatory networks. *Biosystems* 55:129–36.

Wahde, M., and Hertz, J. (2001). Modeling genetic regulatory dynamics in neural development. *J. Comput. Biol.* 8:429–42.

Weaver, D., Workman, C., and Stormo, G. (1999). Modeling regulatory networks with weight matrices. *Pacific Symposium on Biocomputing* 4:122–123.[11]

Wessels, L. F. A., Van Someren, E. P., and Reinders, M. J. T. (2001) A Comparison of genetic network models. *Pacific Symposium on Biocomputing* 6:508–519.[12]

Yeang, C. H., Ramaswamy, S., Tamayo, P., Mukherjee, S., Rifkin, R. M., Angelo, M., Reich, M., Lander, E., Mesirov, J., and Golub, T. (2001). Molecular classification of multiple tumor types. *Bioinformatics* 17(Suppl 1):S316–S322.

---

[9] Available online at http://psb.stanford.edu

[10] Available online at http://psb.stanford.edu

[11] Available online at http://psb.stanford.edu

[12] Available online at http://psb.stanford.edu

8

Molecular Classifiers

If you have two cancer subtypes and you run one chip on each of them, can you then use the chip to classify the cancers into the two subtypes? With 6000 genes or so, easily. You can pick any gene that is expressed in one subtype and not in the other and use that to classify the two subtypes.

What if you have several cancer tissue specimens from one subtype and several specimens from the other subtype? The problem becomes only slightly more difficult. You now need to look for genes that all the specimens from one subtype have in common and are absent in all the specimens from the other subtype.

The problem with this method is that you have just selected genes to fit your data—you have not extracted a *general* method that will *classify any specimen of one of the subtypes that you are presented with after building your classifier*.

In order to build a *general* method, you have to observe several basic rules:

- Avoid overfitting data. Use fewer estimated parameters (degrees of freedom) than the number of specimens that you are building your model on.

- Validate your method by testing it on an independent data set that was not used for building the model. (If your data set is very small, you can use *cross-validation* where you subdivide your data set into test and training several times. If you have ten examples, there are ten ways in which to split the data into a training set of nine and a test set of one. That is called a tenfold cross-validation).

8.1 CLASSIFICATION SCHEMES

8.1.1 Nearest Neighbor

The simplest form of classifier is called a nearest neighbor classifier (Section 12.4.1.6) (Dudoit, et al., 2000; Fix and Hodges, 1951). The general form uses k nearest neighbors and proceeds as follows: (1) for each patient, find the k nearest neighbors according to the distance metric you use; (2) predict the class by majority vote, that is, the class that is most common among the k neighbors. If you use only odd values of k you avoid the situation of a vote tie. Otherwise, vote ties can be broken by a random generator. The value of k can be chosen by cross-validation to minimize the prediction error on a labeled test set.

If the classes are well separated in an initial principal component analysis (Section 4.4) or clustering, nearest neighbor classification will work well. If the classes are not separable by principal component analysis, it may be necessary to use more advanced classification methods, such as neural networks or support vector machines.

8.1.2 Neural Networks

If the number of examples is sufficiently high (between 50 and 100), it is possible to use a more advanced form of classification. Neural networks (Section 12.4.1.7) simulate some of the logic that lies beneath the way in which brain neurons communicate with each other to process information. Neural networks *learn* by adjusting the strengths of connections between them. In computer-simulated artificial neural networks, an algorithm is available for learning based on a learning set that is presented to the software. The neural network consists of an input layer where examples are presented, and an output layer where the answer, or classification category, is output. There can be one or more hidden layers in between the input and output layer.

To keep the number of adjustable parameters in the neural network as small as possible, it is necessary to reduce the dimensionality of array data before presenting it to the network. Khan, et al., (2001) used principal component analysis and presented only the most important principal components to the neural network input layer. They then used an ensemble of cross-validated neural networks to predict the cancer class of patients.

8.1.3 Support Vector Machine

Another type of classifier is the support vector machine (Brown, et al., 2000), a machine learning approach. As it is already designed to work with vectors, it is well suited to the dimensionality of array data.

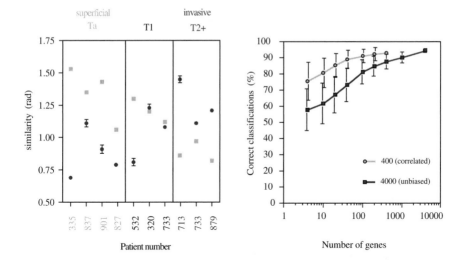

Fig. 8.1 Classifier of bladder cancers based on expression array. Left: Vector angle between patient and reference pool. Each three-digit number on the bottom refers to a patient. The angle between that patient and the two reference pools (squares, Ta pool; circles T2 pool) is indicated. The angle is always smallest (the similarity is greatest) to the pool with the same type of cancer. The intermediate type, T1, for which there is no reference pool, is sometimes more similar to one reference, sometimes more similar to another reference. Error bars have been added to show variation due to choice of reference pool, of which several were available. Right: the performance of the classifier as a function of the number of genes used for classification. Top curve: genes chosen among those 400 genes maximally covarying with the disease. Bottom curve: genes chosen at random from all 4000 genes detected as present in at least one patient. (Christopher Workman based on data from Thykjaer, et al., (2001). See color plate.)

8.2 EXAMPLE I: CLASSIFICATION OF CANCER SUBTYPES

As an example, our lab was faced with the problem of building a classifier that could categorize a bladder cancer as superficial or invasive based on a DNA chip test of a biopsy from the patient (Thykjaer, et al., 2001). We only had biopsies from 10 patients. We decided to use a model without any estimated parameters at all. We simply measured the angle between the vector of all gene expression levels for each patient and the vectors of two reference samples of pools of superficial and invasive cancer. The angle was always smallest to the correct pool, because the vector angle distance was smallest between samples from the same subtype (Figure 8.1). This approach was identical to the k nearest neighbor method with $k = 1$ and vector angle as distance measure. Since we had used no parameters to estimate this classifier, we expected it to be general. It was. When we received four new patient

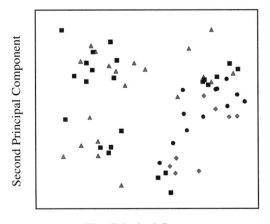

First Principal Component

Fig. 8.2 Principal component analysis of 63 small, round blue cell tumors. Different symbols are used for each of the four categories as determined by classical diagnostics tests. (See color plate.)

samples from our collaborators, they were correctly classified as well. The method was *validated*, although on a very small number of patients.

8.3   EXAMPLE II: CLASSIFICATION OF CANCER SUBTYPES

Khan, et al. (2001) have classified small, round blue cell tumors into four classes using expression profiling and kindly made their data available on the World Wide Web[1]. We can test some of the classifiers mentioned in this chapter on their data.

First, we can try a k nearest neighbor classifier (see Section 12.4.1.6 for details). Using the full data set of 2000 genes, and defining the nearest neighbors in the space of 63 tumors by Euclidean distance, a $k = 3$ nearest neighbor classifier classifies 61 of the 63 tumors correctly in a leave-one-out cross-validation (each of the 63 tumors is classified in turn, using the remaining 62 tumors as a reference set).

We can also train a neural network (see Section 12.4.1.7 for details) to classify tumors into four classes based on principal components. Twenty feed-forward neural network are trained on 62 tumors, and used to predict the class of the 63rd tumor based on a committee vote among the twenty networks (this is a leave-one-out cross-validation). Figure 8.2 shows the first

[1] http://www.thep.lu.se/pub/Preprints/01/lu_tp_01_06_supp.html

two principal components from a principal component analysis. The first ten principal components from a principal component analysis are used as the input for the neural network for each tumor, and four neurons are used for output, one for each category. Interestingly, two of the classes are best predicted with no hidden neurons, and the other two classes are best predicted with a hidden layer of two neurons. Using this setup, the neural networks classify 62 of the 63 tumors correctly. But of course, these numbers have to be validated on an independent, blind set as done by Khan, et al. (2001).

8.4 SUMMARY

The most important points in building a classifier are these:

- Collect as many examples as possible and divide them into a training set and a test set.

- Use as simple a classification method as possible with as few adjustable (learnable) parameters as possible. Advanced methods (neural networks and support vector machines) require more examples for training than nearest-neighbor methods.

- Test the performance of your classifier on the independent test set.

8.5 FURTHER READING

Thykjaer, T., Workman, C., Kruhøffer, M., Demtröder, K., Wolf, H., Andersen, L. D., Frederiksen, C. M., Knudsen, S., and Ørntoft, T. F. (2001). Identification of gene expression patterns in superficial and invasive human bladder cancer. *Cancer Research* 61:2492–2499.

Class discovery and classification

Brown, M. P. S., Grundy, W. N., Lin, D., Cristianini, N., Sugnet, C. W., Furey, T. S., Ares, M., and Haussler, D. (2000). Knowledge-based analysis of microarray gene expression data by using support vector machines *Proc. Natl. Acad. Sci. USA* 97:262–267.

Dudoit, S., Fridlyand, J., and Speed, T. P. (2000). Comparison of discrimination methods for the classification of tumors using gene expression data. Technical report #576, June 2000.[2]

---

[2] Available at http://www.stat.berkeley.edu/tech-reports/index.html

Fix, E., and Hodges, J. (1951). Discriminatory analysis, nonparametric discrimination: Consistency properties. Technical report, Randolph Field, Texas: USAF School of Aviation Medicine.

von Heydebreck, A., Huber, W., Poustka, A., and Vingron, M. (2001). Identifying splits with clear separation: A new class discovery method for gene expression data. *Bioinformatics* 17(Suppl 1):S107–S114.

Hastie, T., Tibshirani, R., Botstein, D., and Brown, P. (2001). Supervised harvesting of expression trees. *Genome Biol.* 2:RESEARCH0003.

Khan, J., Wei, J. S., Ringner, M., Saal, L. H., Ladanyi, M., Westermann, F., Berthold, F., Schwab, M., Antonescu, C. R., Peterson, C., and Meltzer, P. S. (2001). Classification and diagnostic prediction of cancers using gene expression profiling and artificial neural networks. *Nature Genetics* 7:673–679.

Park, P.J., Pagano, M., and Bonetti, M. (2001). A nonparametric scoring algorithm for identifying informative genes from microarray Data. *Pacific Symposium on Biocomputing* 6:52–63.[3]

Xiong, M., Jin, L., Li, W., and Boerwinkle, E. (2000). Computational methods for gene expression-based tumor classification. *Biotechniques* 29:1264–1268.

Yeang, C. H., Ramaswamy, S., Tamayo, P., Mukherjee, S., Rifkin, R. M., Angelo, M., Reich, M., Lander, E., Mesirov, J., and Golub, T. (2001). Molecular classification of multiple tumor types. *Bioinformatics* 17(Suppl 1):S316-S322.

---

[3] Available online at http://psb.stanford.edu

9

Selection of Genes for Spotting on Arrays

You may have so much knowledge of the molecular biology in a particular field that you already know the genes that you wish to spot on a custom array. Say you are interested in a family of proteins, such as a particular class of receptors. If you are not sure that you know all the genes that are part of this family you can do a homology search or a Medline search. Both can be performed at the National Center for Biotechnology Information website[1]. The homology search is best performed starting with the amino acid sequence of one of the core family members and then using Psi-Blast (Altschul, et al., 1997) to iteratively expand the family. The Medline search is done using PubMed by formulating keywords that are specific to your query and then seeing how well the resulting papers that are retrieved match those you are interested in. By iterative reformulation of keywords you should be able to get a reasonable overview of the literature within a selected field—particularly when you look at the literature that has been cited by the relevant papers.

Another way of selecting genes for a spotted array is to use a commercial Affymetrix array to identify genes that are of interest to a particular problem. In that case the t-test or ANOVA (Section 3.5) are reliable tools to select relevant genes that differ significantly between conditions.

[1] http://www.ncbi.nlm.nih.gov

9.1 GENE FINDING

No matter what organism you are working with, there is a large fraction of genes that is not yet functionally characterized. They have been predicted either by the existence of a cDNA or EST clone with matching sequence, by a match to a homologous gene in another organism, or by *gene finding* in the genomic sequence. Gene finding uses computer software to predict the structure of genes based on DNA sequence alone (Guigo, et al., 1992). Hopefully, they are marked as ''hypothetical'' genes by the annotator.

For certain purposes, for example, when designing a chip to measure all genes of a new microorganism, you may not be able to rely exclusively on functionally characterized genes and genes identified by homology. To get a better coverage of genes in the organism you may have to include those predicted by gene finding. Then it is important to judge the quality of the gene finding methods and approaches that have been used. While expression analysis may be considered a good method for experimental verification of predicted genes (if you find expression of the predicted gene it confirms the prediction), this method can become a costly verification if there are hundreds of false positive predictions that all have to be tested by synthesis of complementary oligonucleotides. A recent study showed that for *Escherichia coli* the predicted number of 4300 genes probably contains about 500 false positive predictions (Skovgaard, et al., 2001). The most extreme case is the Archaea *Aeropyrum pernix*, where all open reading frames longer than 100 triplets were annotated as genes. Half of these predictions are probably false (Skovgaard, et al., 2001).

Whether you are working with a prokaryote or a eukaryote, You can assess the quality of the gene finding by looking at which methods were used. If the only method used is looking for open reading frames as in the *A. pernix* case cited above, the worst prediction accuracy will result. Better performance is achieved when including codon usage (triplet) statistics or higher-order statistics (6th-order statistics, for example, measure frequencies of hexamers). These frequencies are to some degree specific to the organism (Cole, et al., 1998). Even better performance is obtained when including models for specific signals like splice sites (Brunak, et al., 1990–1991), promoters (Knudsen, 1999; Scherf, et al., 2000), and start codons (Guigo, et al., 1992). Such signals are best combined within hidden Markov models which seem particularly well suited to the sequential nature of gene structure (Borodovski and McIninch, 1993; Krogh, 1997; Burge and Karlin, 1997).

9.2 SELECTION OF REGIONS WITHIN GENES

Once you have the list of genes you wish to spot on the array, the next question is one of cross-hybridization. How can you prevent spotting probes that are complementary to more than one gene? This question is of particular importance if you are working with a gene family with similarities in sequence. There is software available to help search for regions that have least similarity (determined by Blast; Altschul, et al., 1990) to other genes. At our lab we have developed ProbeWiz[2] (Nielsen and Knudsen, 2001), which takes a list of gene identifiers and uses Blast to find regions in those genes that are the least homologous to other genes. It uses a database of the genome from the organism you are working with. Current databases available include: *Homo sapiens, Caenorhabditis elegans, Drosophila melanogaster, Arabidopsis thaliana, Saccharomyces cerevisae,* and *Escherichia coli.*

9.3 SELECTION OF PRIMERS FOR PCR

Once those unique regions have been identified, the probe needs to be designed from this region. It has been customary to design primers that can be used for polymerase chain reaction (PCR) amplification of a probe of desired length. ProbeWiz will suggest such primers if you tell it what length of the probe you prefer and whether you prefer to have the probe as close to the 3' end of your mRNA as possible. It will attempt to select primers whose melting temperature match as much as possible.

9.4 SELECTION OF UNIQUE OLIGOMER PROBES

There is a trend in spotted arrays to improve the array production step by using long oligonucleotides (50 to 70 basepairs) instead of PCR products. Li and Stormo (2001) have run their DNA oligo (50–70 bases) prediction software on a number of complete genomes and made the resulting lists available online.[3]

9.4.1 Example: Finding PCR Primers for Gene AF105374

GenBank accession number AF105374 (Homo sapiens heparan sulfate D-glucosaminyl 3-O-sulfotransferase-2) has been submitted to the web version of ProbeWiz, and the output generated if standard settings are used is given in Figure 9.1.

In addition to suggesting two primers for the PCR amplification, ProbeWiz gives detailed information on each of these primers and their properties, as well as a number of scoring results from the internal weighting process that went into selection of these two primers over others. The latter information may be useful only if you ask for more than one suggestion per gene and if you are comparing different suggestions.

---

[2] Available in a web version at http://www.cbs.dtu.dk/services/DNAarray/probewiz.html
[3] Available at http://ural.wustl.edu/ lif/probe.pl

```
EST ID AF105374
Left primer sequence TGATGATAGATATTATAAGCGACGATG
Right primer sequence AAGTTGTTTTCAGAGACAGTGTTTTC
PCR product size 327
Primer pair penalty 0.6575 (Primer3)
```

	left primer	right primer
Position	1484	1810
Length	27	27
TM	59.8	60.4
GC %	33.3	33.3
Self annealing	6.00	5.00
End stability	8.60	7.30

Penalties:	Weighted	Unweighted
Homology	0	0
Primer quality	65.75	0.657
3'endness	158	158

Fig. 9.1 Output of ProbeWiz server upon submission of the human gene AF105374 (*Homo sapiens* heparan sulfate D-glucosaminyl 3-O-sulfotransferase-2).

9.5 EXPERIMENTAL DESIGN

In the early days of robot spotting, the amount of probe spotted varied so much between slides that it was inadvisable to compare expression of a gene between different slides. Comparing only the red and green channels on a single slide canceled any printing tip effects.

With advances in printing tip technology, however, it is possible to deposit so reproducible amounts of probe on individual slides that it is possible to compare expression between slides, provided that each condition is replicated over several slides so a *t*-test can be used to measure and correct for variation between slides (Workman, et al., 2001).

9.6 FURTHER READING

Workman, C., Jensen, L. J., Jarmer, H., Saxild, H. H., Berka, R., Gautier, L., Nielsen, H. B., Nielsen, C., Brunak, S., and Knudsen, S. (2001). A new non-linear method to reduce variability in DNA microarray experiments. Manuscript submitted.

Primer selection tools

Li, F., and Stormo, G. D. (2001). Selection of optimal DNA oligos for gene expression arrays. *Bioinformatics* 17:1067–1076.

Nielsen, H. B., and Knudsen, S. (2001). Avoiding cross hybridization by choosing nonredundant targets on cDNA arrays. *Bioinformatics* In press.

Varotto, C., Richly, E., Salamini, F., and Leister, D. (2001). GST-PRIME: A genome-wide primer design software for the generation of gene sequence tags. *Nucleic Acids Res.* 29:4373–7.

Gene finding

Altschul, S. F., Gish, W., Miller, W., Myers, E. W., and Lipman, D. J. (1990). Basic local alignment search tool. *J. Mol. Biol.* 215:403–410.[4]

Altschul, S. F., Madden, T. L., Schäffer, A. A., Zhang, J., Zhang, Z., Miller, W., and Lipman, D. J. (1997). Gapped BLAST and PSI-BLAST: A new generation of protein database search programs. *Nucleic Acids Res.* 25:3389–3402.[4]

Borodovsky, M., and McIninch, J. (1993). GeneMark: Parallel gene recognition for both DNA Strands. *Computers & Chemistry* 17:123–133.

Brunak, S., Engelbrecht, J., and Knudsen, S. (1990). Cleaning up gene databases. *Nature* 343:123.

Brunak, S., Engelbrecht, J., and Knudsen, S. (1990). Neural network detects errors in the assignment of mRNA splice sites. *Nucleic Acids Res.* 18:4797–4801.

Brunak, S., Engelbrecht, J., and Knudsen, S. (1991). Prediction of human mRNA donor and acceptor sites from the DNA sequence. *Journal of Molecular Biology* 220:49–65.

Burge C., and Karlin, S. (1997). Prediction of complete gene structures in human genomic DNA. *Journal of Molecular Biology* 268:78–94.

Cole, S. T., et al. (1998). Deciphering the biology of *Mycobacterium tuberculosis* from the complete genome sequence. *Nature* 393:537–544.

Guigo, R., Knudsen, S., Drake, N., and Smith. T. (1992). Prediction of Gene Structure. *Journal of Molecular Biology* 226:141–157.

---

[4] Available at http://www.ncbi.nlm.nih.gov/BLAST/

Knudsen, S. (1999). Promoter2.0: For the recognition of PolII promoter sequences. *Bioinformatics* 15:356–361.[5]

Krogh, A. (1997). Two methods for improving performance of an HMM and their application for gene finding. *Proc. Fifth Int. Conf. on Intelligent Systems for Molecular Biology (ISMB)* Menlo Park, CA: AAAI Press, pp. 179–186.

Scherf, M., Klingenhoff, A., and Werner, T. (2000). Highly specific localization of promoter regions in large genomic sequences by PromoterInspector: A novel context analysis approach. *Journal of Molecular Biology* 297:599–606.

Skovgaard, M., Jensen, L. J., Brunak, S., Ussery, D., and Krogh, A. (2001). On the total number of genes and their length distribution in complete microbial genomes. *Trends Genet.* 17:425–428.

---

[5]Available as a web server at http://www.cbs.dtu.dk/services/Promoter/

10

Limitations of Expression Analysis

For all its strengths, it is important to keep in mind the limitations of expression analysis. First, expression analysis measures only the transcriptome. Important regulation takes place at the level of translation and enzyme activity. Those regulation effects are, at least for now, ignored in any analysis. In fact, significant signal transduction takes place at a protein to protein level. The only effect of such a signal transduction that you can observe in a gene expression experiment is any effect on gene expression that may be at the end of the signal transduction pathway.

Another issue that is largely ignored in current expression analysis is the effect of alternative splicing. To what extent are changes in observed signal from a particular messenger due to alternative splicing rather than due to a change in abundance? Current DNA microarrays for expression analysis have not been designed to distinguish between these two effects, largely because current knowledge of alternative splicing in the transcriptome is so limited.

In theory, the approach of using multiple probes per gene should be able to reveal alternative splicing if probes span an alternative splice junction. Thus, looking for changes in relative probe intensity within a gene might reveal alternative splicing (Hu, et al., 2001) but it does not exclude the possibility of changes in cross-hybridisation for some of the probes within a gene.

Finally, keep in mind that mRNA is an unstable molecule. Messengers are programmed for enzymatic degradation and half-lives of messengers vary considerably. Messengers with very short half-lives may be difficult to extract in reproducible quantity. Thus, any regulation in expression of a gene with a very short half-life may be impossible to detect with statistical significance in a method that relies on reproducibility. Those unstable messengers will simply never pass a *t*-test.

In fact, unless the enzymatic degradation of messengers is stopped immediately after extraction of the sample (for example, by cooling in liquid nitrogen), there is not likely to be any unstable messengers left in the mRNA extraction. Those messengers will never be detected as present in your hybridization even though they may have been present inside the living cell.

For mRNA that has been extracted without proper care to immediately stop all degradation of messengers, all that is left is to analyze any changes in expression of stable messengers.

10.1   RELATIVE VERSUS ABSOLUTE RNA QUANTIFICATION

Most of this book has focused on relative changes in the abundance of a messenger RNA. Absolute quantification is a much harder task. You need to know how well each probe hybridizes to its target before you can use it to deduce anything about the absolute concentration of mRNA in the cell. One approach is to calibrate each probe set using known concentrations of their corresponding mRNA. That is labor intensive if you are working with many different mRNAs.

It is for that reason that most absolute analysis of mRNA has limited itself to the determination of whether a particular mRNA was present or not. Affymetrix in early version of their software made such a call based on empirically determined rules that take into account the number of PM/MM probe pairs which have a positive difference above a certain threshold, the number of PM/MM probe pairs which have a negative difference above a certain threshold, and the average log-ratio of all probe pairs, log(PM/MM). Schadt et al., (2000) have developed a statistics-based approach to the presence/absence determination.

Still, all you can say is whether a certain messenger RNA is above detection threshold or not. If it is significantly above, you may be able to say with high confidence that it is present. But if you don't detect a messenger, you cannot rule out that it is expressed below detection limit.

10.2   FURTHER READING

Detection of alternative splicing with Affymetrix chips

Hu, G. K., Madore, S. J., Moldover, B., Jatkoe, T., Balaban, D., Thomas, J., and Wang, Y. (2001). Predicting splice variant from DNA chip expression data. *Genome Research* 11:1237–1245.

Statistical detection of presence with Affymetrix arrays

Schadt, E. E., Li, C., Su, C., and Wong, W. H. (2000). Analyzing high-density oligonucleotide gene expression array data. *J. Cell. BioChem.* 80:192–201.

11

Genotyping Chips

Up to this point this text has covered analysis of expression data. Chips for genotyping are also available. Instead of measuring mRNA, they measure DNA. For example, a p53 chip is available from Affymetrix for detecting mutations in the DNA of human p53 tumor supressor gene. It does so with overlapping oligos that each are complementary to 20 base pairs of the *TP53* gene (Figure 11.1). Each oligo is present in 5 versions: the central nucleotide is either an A, C, G, T or is absent (a 1 bp deletion). Only one of these five oligos corresponds to the wild type (nonmutant) version of the *TP53* gene.

The PCR amplified, fragmented, fluorescently labeled DNA from a patient sample will hybridize to the complementary oligo for each position in the gene and it is then possible to "read" the sequence of the entire gene and determine whether it is equal to the wild type or not. There are still limitations of the accuracy of this determinination (Ahrendt, et al., 1999; Wikman, et al., 2000). Our lab has been working on neural network-based software that will improve the determination.

11.1 EXAMPLE: NEURAL NETWORKS FOR GENECHIP PREDICTION

Neural networks can be trained to predict DNA sequence based on the hybridization intensities measured on a chip designed for a specific gene (Spicker, et al., 2001). It requires a large training set of DNA with accurately determined sequence.

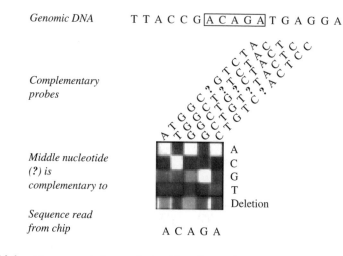

Fig. 11.1 p53 gene mutation analysis chip. Shown is a small area of the chip surface containing 25 different oligos designed to read the sequence of 5 nucleotides in the gene. The oligo probes (reduced to length 11 for clarity) for each position differ in their middle nucleotide (?) to account for the four possible bases as well the possibility of a one nucleotide deletion (right). By reading the intensity of the fluorescence at each oligo it is possible to read the DNA sequence. (Experiment by Jeppe Spicker.)

Neural networks simulate some of the logic that lies beneath the way in which brain neurons communicate with each other to process information. Neural networks *learn* by adjusting the strengths of connections between them. In computer-simulated artificial neural networks, algorithms are available for learning based on a learning set that is presented to the software. In our case (Figure 11.2), the neural network consisted of an input layer which was presented with the measured fluorescence intensities for each of the ten probes used to detect each position: five with a central nucleotide of A, C, G, T, or deletion on the sense strand, and five with a central nucleotide of A, C, G, T, or deletion on the antisense strand. The output layer is then used to predict which nucleotide is present at the given position in the *TP53* gene. In between the input and output layers is a hidden layer which performs data processing. Such a neural network can then be trained by presenting examples matching input and output — *TP53* genes where the sequence has been determined by alternative means. When trained, it can predict the sequence based on a chip. This method is, however, sensitive to inhomogeneous samples where the sequence is ambiguous (Wikman, et al., 2000), because it is a mixture of two alleles.

As an interesting aside, our neural network discovered an error in the labeling of input data during training. There were a few positions in the patient material that the neural network could not learn. Closer inspection

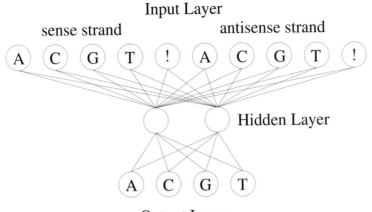

Fig. 11.2 Neural network. Shown are the units in the input layer, hidden layer, and output layer. All units in one layer are connected with adjustable strengths to all units in the adjacent layer. (Figure by Jeppe Spicker.)

revealed that it had detected known sites of polymorphisms that we had failed to label correctly in the training data (Spicker, et al., 2001).

11.2 FURTHER READING

Ahrendt, S. A., Halachmi, S., Chow, J. T., Wu, L., Halachmi, N., Yang, S. C., Wehage, S., Jen, J., and Sidransky, D. (1999). Rapid p53 sequence analysis in primary lung cancer using an oligonucleotide probe array. *Proc. Natl. Acad. Sci. USA* 96:7382–7.

Hacia, J. G. (1999). Resequencing and mutational analysis using oligonucleotide microarrays (Review). *Nat. Genet.* 21(1 Suppl):42–7.

Spicker, J. S., Wikman, F, Lu M. L., Cordon-Cardo C., Ørntoft, T. F., Brunak, S., and Knudsen, S. (2001). Neural network trained to predict TP53 gene sequence based on p53. Manuscript submitted.

Wikman, F. P., Lu, M. L., Thykjaer, T., Olesen, S. H., Andersen, L. D., Cordon-Cardo, C., and Ørntoft, T. F. (2000). Evaluation of the performance of a p53 sequencing microarray chip using 140 previously sequenced bladder tumor samples. *Clin. Chem.* 46:1555–61.

12

Software Issues and Data Formats

In the future, sophisticated statistical, computational, and database methods may be as commonplace in Molecular Biology and Genetics as recombinant DNA is today.

—Pearson, 2001

Software for array data analysis is a difficult issue. There are commercial software solutions and there are noncommercial, public domain software solutions. In general, the commercial software lacks flexibility or sophistication and the non-commercial software lacks user-friendliness or stability.

Take Microsoft's Excel spreadsheet, for example. Many biologists use it for their array data processing, and indeed it can perform many of the statistical and numerical analysis methods described in this book. But there are pitfalls. First of all, commercial software packages can make assumptions about your data without asking you. As a scientist, you don't like to lose control over your calculations in that way. Second, large spreadsheets can become unwieldy and time-consuming to process and software stability can become an issue. Third, complicated operations require macro programming, and there are other, more flexible environments for programming.

You can perform the same types of analysis using non-commerical software on a Unix/Linux system where the limits are few and the analysis is rigorous. But it may require you to change the format of your data as you go from one software program to the next. Such format conversions can be performed in Excel, but then we are back to square one.

The best solution is to learn a single programming language like *Awk* or *Perl* that will allow you to manipulate your data in any way you want before presenting it for analysis in a software package. It is then that you will experience the unlimited power of the Unix/Linux system when combined with a programming language. Considering that the Linux system is more or less free, comes with Perl and Awk preinstalled, and runs on just about any PC, this really is an attractive solution.

Awk is available on all Unix/Linux systems. A GNU version can be downloaded from www.gnu.org/directory/gawk.html. A good book for learning and using Awk is *The AWK Programming Language*, by A. V. Aho, B. W. Kernigan, and P. J. Weinberger (ISBN 0-201-07981-X).

Perl is available on most Unix/Linux systems and can be downloaded for free for most systems at www.cpan.org (see also www.perl.org). A good book for learning perl is *Learning Perl*, by R. L. Schwartz and T. Christiansen (ISBN 1-56592-284-0).

Patrick Brown's lab at Stanford University keeps a web page with links to various public domain software for array analysis for different computer platforms.[1]

12.1   STANDARDIZATION EFFORTS

There are several efforts underway to standardize file formats as well as the description of array data and underlying experiments. Such a standard would be useful for the construction of databases and for the exchange of data between databases and between software packages.

- A number of companies and entities have agreed on a Gene Expression Markup Language (GEML[2]) which is based on the XML standard used on the World Wide Web. It uses tags, known from html, to describe array data and information. A White Paper describing GEML is available[2].

- The GATC Consortium[3] founded by Affymetrix and Molecular Dynamics has proposed a standard for expression array database structure.

- The Microarray Gene Expression Database Group[4] has proposed "Minimum Information About a Microarray Experiment" for use in databases.

[1] brownlab.stanford.edu
[2] http://www.geml.org
[3] http://www.gatconsortium.org
[4] www.mged.org

12.2 STANDARD FILE FORMAT

The following standard file format is convenient for handling and analyzing expression data and will work with most software. Information about the experiment, the array technology, and the scaling is not included in this format. The file should be one line of tab-delimited fields for each gene (probe set):

Field 1: gene ID or GenBank accession number

Field 2: (optional) text describing function of gene

Field 3 and up: intensity values for each experiment including control, or logfold change for each experiment relative to the control

The file should be in text format, and not in proprietary, inaccessible formats that some commercial software houses are fond of using. The data should be normalized by whatever method you have chosen for your particular problem.

12.2.1 Example: Small Scripts in Awk

Once you have a file in this format you can use standard Unix/Linux tools to process it. Here's an Awk script for converting a text file in the above format from AffyFold to Logfold (see Section 3.4):

```
awk -F"\t" '{printf"%s\t%s\t",$1,$2; \
  for(i = 3; i <= NF; i + +) \
    if($i < 0) \
      printf"%3.2f\t", log(-1/$i); \
    else \
      printf"%3.2f\t", log($i); \
  printf"\n"}' infile >outfile
```

The backslashes are included to allow you to enter input that extends over several lines on one command line.

Here's an Awk script for calculating Logfold based on AvgDiffs (the control is in the third column; all values less than 20 are set to 20):

```
awk -F"\t" '{printf"%s\t%s\t",$1,$2; \
  if($3 < 20) \
    ctrl = 20; \
  else \
    ctrl = $3; \
  for(i = 4; i <= NF; i + +) \
    if($i < 20) \
      printf"%3.2f ",log(20/ctrl); \
    else \
      printf"%3.2f ",log($i/ctrl) \
  }' infile >outfile
```

You can simply type this program on the command line of any Unix/Linux system and it will perform the task. The advantage of Awk in this context is that it is specially designed to handle data in text table format.

12.3   SOFTWARE FOR CLUSTERING

Once you have run either of these programs and have your data in a tab-delimited text format of Logfold values or Average Difference values (see file format specification listed above), you can input it to the ClustArray software, which can do further data analysis for you:

- Rank genes based on vector length, covariance to disease progression, etc.

- Perform normalizations

- Cluster using a selection of different clustering algorithms and distance metrics

- Produce tree files to be visualized with drawtree or similar program

- Produce a distance matrix

- Visualize gene clusters with graphic Postscript file

If ClustArray is installed on your system, type ClustArray -h to get a full list of all the options.

A web interface to this software is available for the most commonly used purposes.[5] ClustArray is sensitive to the format of the input data, which must be tab separated and start with a line beginning with COLUMN_LABELS. ClustArray is available for Unix/Linux platforms.[5]

Other available software for clustering:

- Cluster. A widely used and user friendly software by Michael Eisen.[6] It is for Microsoft Windows only.

- GeneCluster. From the Whitehead/MIT Genome Center. Available for PC, Mac and Unix.[7]

- Expression Profiler. A set of tools from the European Bioinformatics Institute that perform clustering, pattern discovery, and gene ontology browsing. Runs in a web server version and is also available for download in a Linux version.[8]

---

[5]http://www.cbs.dtu.dk/services/DNAarray/
[6]http://rana.lbl.gov
[7]http://www-genome.wi.mit.edu/MPR/
[8]http://ep.ebi.ac.uk

Table 12.1 Expression readings of four genes in six patients.

Gene	N_1	N_2	A_1	A_2	B_1	B_2
			Patient			
a	90	110	190	210	290	310
b	190	210	390	410	590	610
c	90	110	110	90	120	80
d	200	100	400	90	600	200

12.3.1 Example: Clustering with ClustArray

We will use a small standard example for clarity (Table 12.1). This can be converted into the format required by ClustArray with the following awk script:

```
awk 'BEGIN{printf"COLUMN_LABELS\t"} \
    {for(i = 1; i < NF; i + +) printf"%s\t",$i; printf"%s\n",$NF}' \
    example >exampleCE
```

Now we can use ClustArray to hierarchically cluster genes based on vector angle distance:

```
ClustArray -f exampleCE -m exampleCE.mtx \
    -t exampleCE.new -Ci 1 -Ca 0 -Cd 2  -Cm 2 -Cc 1
```

The software has produced two files: exampleCE.mtx contains the distance matrix and exampleCE.new contains a Newick description of the resulting tree. Most treeviewers (Phylip's drawgram, TreeView, etc.) can read this tree format and draw the tree for you. The resulting drawing is shown in Figure 5.6.

12.4 SOFTWARE FOR STATISTICAL ANALYSIS

Microsoft Excel has several statistics functions built in, but an even better choice is the free, public-domain R package which can be downloaded for Unix/Linux, Mac, and Windows systems.[9] This software can be used for *t*-test, ANOVA, principal component analysis, clustering, classification, neural networks, and much more.

12.4.1 Example: Statistical Analysis with R

In this section we will use a small standard example for clarity (Table 12.2). There are four genes, each measured in six patients, which fall into three

[9]http://www.r-project.org

Table 12.2   Expression readings of four genes in six patients.

| Gene | Patient | | | | | |
	N_1	N_2	A_1	A_2	B_1	B_2
a	90	110	190	210	290	310
b	190	210	390	410	590	610
c	90	110	110	90	120	80
d	200	100	400	90	600	200

categories: normal (N), disease stage A, and disease stage B. That means that each category has been *replicated* once. The data follow the standard format except that there is no function description.

For some analyses the patient label will confuse the software, but it is easy to remove. If the example above is in the file named ''example,'' we can create a new file named ''example2'' with the patient labels removed:

> **awk** $'NR > 1'$ example >example2

Next you boot up R (by typing the letter R at the prompt) and read in the file:

```
dataf  <- read.table("example2")
dataf[,1] <- as.character(dataf[,1])
```

12.4.1.1   The *t*-test   Here is how you would perform a *t*-test to see if genes differ significantly between patient category A and patient category B:

```
# load library for t-test:
  library(ctest)
# t-test function:
  get.pval.ttest <- function(dataf,index1,index2,
    datafilter=as.numeric){
    f <- function(i) {
      return(t.test(datafilter(dataf[i,index1]),
        datafilter(dataf[i,index2]))$p.value)
    }
    return(sapply(1:length(dataf[,1]),f))
  }
# call function with our data:
  pVal.ttest <- get.pval.ttest(dataf,4:5,6:7)
# print results on screen (only for a small dataset like this)
  print(cbind(dataf,pVal.ttest))
         V1  V2  V3  V4  V5  V6  V7 pVal.ttest
  1 gene_a  90 110 190 210 290 310 0.01941932
  2 gene_b 190 210 390 410 590 610 0.00496281
  3 gene_c  90 110 110  90 120  80 1.00000000
  4 gene_d 200 100 400  90 600 200 0.60590011
# sort on P-value and write to file:
  orders <- order(pVal.ttest)
  ordered.data <- cbind(dataf[orders,],pVal.ttest[orders])
  write(t(as.matrix(ordered.data)),
    ncolumns=length(dataf)+1,
```

```
file = "ttest.out")
q(save="no")
```

The default call to the t.test function assumes unequal variance between the two populations (Welch's *t*-test).

12.4.1.2 Wilcoxon You can perform the Wilcoxon test instead of the *t*-test by replacing the call to the t.test function above with a call to the wilcox.test function.

12.4.1.3 ANOVA Here is how you would perform an ANOVA on the example to test for genes that differ significantly in at least one of the three categories N, A, and B:

```
# Specify categories and columns holding AvgDiff data:
  Categories <- as.factor(c("0","0","A","A","B","B"))
  indexAvgDiff <- 2:7
# ANOVA function:
  get.pval.anova <-
  function(dataf,indexAll,Categories,
    datafilter=as.numeric){
    Categories <- as.factor(Categories)
    f <- function(i) {
      return(summary(
        aov(datafilter(dataf[i,indexAll]) ~ Categories)
      )
      [[1]][4:5][[2]][1])
    }
    return(sapply(1:length(dataf[,1]),f))
  }
# call the function with our data:
  pVal.anova <- get.pval.anova(dataf,indexAvgDiff,Categories)
# print results on screen (only for a small dataset like this)
  print(cbind(dataf,pVal.anova))
          V1   V2   V3   V4   V5   V6   V7    pVal.anova
  1 gene_a  90  110  190  210  290  310  0.0017965439
  2 gene_b 190  210  390  410  590  610  0.0002283540
  3 gene_c  90  110  110   90  120   80  1.0000000000
  4 gene_d 200  100  400   90  600  200  0.5560965577
# sort on P-value and write results to file:
  orders <- order(pVal.anova)
  ordered.data <- cbind(dataf[orders,],pVal.anova[orders])
  write(t(as.matrix(ordered.data)),
   ncolumns=length(dataf)+1,
   file = "ANOVA.out")
  q(save="no")
```

In general, it is best to perform statistical tests on the raw AvgDiff values (including negative AvgDiff values) instead of fold change values where AvgDiffs below 20 or below 0 have been clipped. If you are working with spotted arrays where there is much variation between slides it may be better to perform the statistical test on the fold change derived from the red and green channels of each slide. Try to perform the statistical test on both absolute values and fold change and compare the results. Baldi and Long (2001) advocate log-transformation of data before statistical analysis.

12.4.1.4 PCA   You can perform a principal component analysis of the same data as follows (note that the first row of the data must contain labels of all patients, but no label for the gene identifier):

```
library(mva)
dataf  <- read.table("example")
pca <- princomp(dataf)
summary(pca)
plot(pca)
biplot(pca)
```

This will produce the plot shown in Figure 4.2 in Section 4.1. The plot shown in Figure 4.3 was produced by transposing the data matrix with the command t(dataf), but PCA may not work for large transposed matrices.

12.4.1.5 Correspondence analysis   You can perform a correspondence analysis much the same way as you perform a principal component analysis:

```
library(MASS)
library(mva)
dataf  <- read.table("example",header=T)
cs <- corresp(dataf,nf=2)
plot(cs)
```

12.4.1.6 Classification   You can perform k nearest neighbor classification with cross-validation using builtin functions in the standard distribution of R. This is the calculation that was performed for the classificer in example 8.3:

```
library(class)
library(mva)
dataf  <- read.table("datafile",header=T)
knn.targets <- factor( c(rep("E", 23), rep("B", 8),
                         rep("N", 12), rep("R", 20)) )
knn.cv(dataf, knn.targets, k=3, prob=TRUE)
E E E E E E E E E E E E E E E E E E E E E E E
B B B B B B B N N N N N N N N N N N N R R R N
R R R R R E R R R R R R R R R
```

The predicted classes (last two lines) differ from those assigned (knn.targets) for two tumors, so this classification is 97% correct.

12.4.1.7 Neural networks   The R package has a function for training a feed-forward neural network and using the trained network to predict the class of unlabeled samples. These calculations were performed for the neural network results shown in example 8.3 with leave-one-out cross-validation:

```
library(class)
library(mva)
pca <- princomp(dataf,cor=TRUE,scores=TRUE)
canc <- pca$scores[,1:10]
targets <- class.ind( c(rep("E", 23), rep("B", 8),
```

```
                              rep("N", 12), rep("R", 20)) )
f <- function(samp) {
    b <- c(0,0,0,0)
    for(i in 1:20) {
        trainednet <- nnet(canc[-samp,], targets[-samp,],
                size=2,skip=FALSE,trace=FALSE, maxit=300)
        a <- max.col(predict.nnet(trainednet, canc[samp,]))
        b[a] <- b[a] + 1
    }
    print(b)
}
lapply(1:63,f)
```

12.4.2 The affyR Software Package

In our lab we have developed an R package[10] (Gautier and Knudsen, 2001) that, in addition to the above-mentioned statistical analyses, can replace Affymetrix GeneChip software by reading CEL files directly and calculating expression values using the Li-Wong method as well as other methods. A number of other R packages for gene expression analysis have been developed[11]

12.4.3 Commercial Statistics Packages

The commercial statistics packages SAS, SPSS, and S-PLUS include the statistical functions described in this section as well.

12.5 SUMMARY

Unix or Linux operating systems offer the best solution for data analysis. Linux runs on any PC and opens up a world of free downloadable analysis software packages that you can employ in your data analysis if you learn a simple programming language that will allow you to structure your data for each analysis. Perl and Awk are such simple-to-use all-round programming languages that can be downloaded for free (or are included in the Linux distribution). With such a setup your possibilities for data mining and discovery are limited only by your imagination (and, perhaps, by your programming skills).

---

[10]http://www.cbs.dtu.dk/biotools/affyR/

[11]http://users.ox.ac.uk/ strimmer/rexpress.html

12.6   FURTHER READING

Baldi, P., and Long, A. D. (2001). A Bayesian framework for the analysis of microarray expression data: Regularized t-test and statistical inferences of gene changes. *Bioinformatics* 17:509–519.[12]

Pearson, W. R. (2001). Training for Bioinformatics and Computational Biology (Editorial). *Bioinformatics* 17:761–762.

Awk

Aho, A. V., Kernigan, B. W., and Weinberger, P.J. (1988). *The AWK Programming Language*. Reading, MA: Addison-Wesley.

Perl

Schwartz, R. L. and Phoenix, T. (2001). *Learning Perl*, 3rd ed. Sebastopol, CA: O'Reilly & Associates

Christiansen, T., Torkington, N., and Wall, L. (1998). *Perl Cookbook*, 1st ed. Sebastopol, CA: O'Reilly & Associates.

Wall, L., Christiansen, T., and Orwant, J. (2000). Programming Perl, 3rd ed. Sebastopol, CA: O'Reilly & Associates.

R

Online manuals available at http://www.r-project.org.

Venables, W. N. and Ripley, B. D. (1999). *Modern applied statistics with S-PLUS*, 3rd ed. New York, NY: Springer.

Unix/Linux

Bourne, S. (1983). *The UNIX System*. Reading, MA: Addison-Wesley.

Other public domain software

Patrick Brown's lab at Stanford University keeps a web page with links to various public domain software for array analysis for different computer platforms.[13]

Gautier, L., and Knudsen, S. (2001). An R software environment for analysis of oligonucletide array data. Manuscript in preparation.

---

[12] Accompanying web page at http://128.200.5.223/CyberT/
[13] http://brownlab.stanford.edu

13

Commercial Software Packages

The previous chapter introduced the Unix/Linux platform and some free software programs that can perform all the analyses described in this book. Such an environment is most suited for bioinformatics work because it has unlimited flexibility. No DNA array analysis tasks are identical and you need to tailor each analysis to the problem at hand. With Linux and some simple programming you can do that.

However, not every biologist wants to deal with Linux and programming. There are alternatives. Complete software packages for performing analysis of DNA microarray data are available for Microsoft Windows platforms and other platforms. I will briefly mention some of them and describe the functionality that they currently include.

- **Affymetrix Data Mining Tool**
  This software offers statistical analysis as well as clustering and visualization of Affymetrix GeneChip data. It is integrated with LIMS to offer data management of large volumes of chip data. The price is targeted at large pharmaceutical companies.
  http://www.affymetrix.com

- **Affymetrix NetAffx**
  An online resource freely accessible to Affymetrix customers, it links probes on Affymetrix GeneChips to public and proprietary databases, allowing users to integrate data analysis from their expression experiments.
  http://www.netaffx.com

- **Biomax Gene Expression Analysis Suite**
  Clusters genes and links genes in selected clusters to metabolic pathways. Protein interaction networks of co-expressed genes are identified from a database of protein interactions for the respective organism.
  http://www.biomax.de

- **GeneData Expressionist**
  A software system for organizationwide analysis of expression data from a variety of platforms including Affymetrix and spotted cDNA arrays and filters. Expressionist performs background subtraction, scaling, statistical tests, clustering, data visualization, and promoter searching. The software is targeted at biotech industries.
  http://www.genedata.com/

- **Informax Xpression NTI**
  Imports expression data in a variety of formats, performs normalization, clustering, and graphical representation of data. Informax also offer ArrayPro Analyzer to process spotted array images to expression values.
  http://www.informaxinc.com

- **Invitrogen Corporation ResGen Pathways**
  A comprehensive set of tools for the analysis of differential gene expression using microarray data. Image analysis, statistical analysis, clustering, pathway analysis and linking to databases.
  http://pathways.resgen.com/

- **Lion Bioscience arraySCOUT**
  Enterprisewide expression data analysis designed to handle large volumes of expression data. The software includes statistical analysis, clustering, and visualization tools. Includes database of gene annotation.
  http://www.lion-bioscience.com/

- **Rosetta Resolver Gene Expression Data Analysis System**
  The Rosetta Resolver is an enterprisewide gene expression data analysis system that combines analysis software, a high capacity database, and high-performance server hardware to enable users to store, retrieve, and analyze large volumes of gene expression data. It is accessed from PC clients. The software includes statistical analysis, five different clustering algorithms, and numerous visualization tools. The price is targeted at very large pharmaceutical companies.
  http://www.rosettabio.com//

- **Silicon Genetics GeneSpring**
  Performs clustering, multiple visualizations, and annotation of expression data from multiple sources. It includes a regulatory sequence

search algorithm. The price is targeted at smaller research groups.
http://www.sigenetics.com

- **Spotfire**
  A multipurpose tool for analyzing and visualizing data.
  http://www.spotfire.com/

Most of these software packages use Microsoft Windows. GeneSpring is available in a Macintosh version as well.

14

Bibliography

Affymetrix (1999). *GeneChip Analysis Suite*. User Guide, version 3.3.

Affymetrix (2000). *GeneChip Expression Analysis*. Technical Manual.

Ahrendt, S. A., Halachmi, S., Chow, J. T., Wu, L., Halachmi, N., Yang, S. C., Wehage, S., Jen, J., and Sidransky, D. (1999). Rapid p53 sequence analysis in primary lung cancer using an oligonucleotide probe array. *Proc. Natl. Acad. Sci. USA* 96:7382–7.

Akutsu, T., Miyano, S., and Kuhara, S. (1999). Identification of genetic networks from a small number of gene expression patterns under the boolean network model. *Pacific Symposium on Biocomputing* 4:17–28.[1]

Alter, O., Brown, P. O., and Botstein, D. (2000). Singular value decomposition for genome-wide expression data processing and modeling. *Proc. Natl. Acad. Sci. USA* 97:10101–6.

Altschul, S. F., Gish, W., Miller, W., Myers, E. W., and Lipman, D. J. (1990). Basic local alignment search tool. *J. Mol. Biol.* 215:403–410.[2]

Altschul, S. F., Madden, T. L., Schäffer, A. A., Zhang, J., Zhang, Z., Miller, W., and Lipman, D. J. (1997). Gapped BLAST and PSI-BLAST: A

---

[1] Available online at http://psb.stanford.edu
[2] Available at http://www.ncbi.nlm.nih.gov/BLAST/

new generation of protein database search programs. *Nucleic Acids Res.* 25:3389–3402.[2]

Audic, S., and Claverie, J. M. (1997). The significance of digital gene expression profiles. *Genome Res.* 7:986–995.

Baldi, P., and Long, A. D. (2001). A Bayesian framework for the analysis of microarray expression data: Regularized t-test and statistical inferences of gene changes. *Bioinformatics* 17:509–519.[3]

Baugh, L. R., Hill, A. A., Brown, E. L., and Hunter, C. P. (2001). Quantitative analysis of mRNA amplification by in vitro transcription. *Nucleic Acids Research* 29:E29.

Bender, R., and Lange, S. (2001). Adjusting for multiple testing—when and how? *Journal of Clinical Epidemiology* 54:343–349.

Birnbaum, K., Benfey, P. N., and Shasha, D. E. (2001). Cis element/transcription factor analysis (cis/TF): A method for discovering transcription factor/cis element relationships. *Genome Res.* 11:1567–1573.

Borodovsky, M., and McIninch, J. (1993). GeneMark: Parallel gene recognition for both DNA Strands. *Computers & Chemistry* 17:123–133.

Bourne, S. (1983). *The UNIX System.* Reading, MA: Addison-Wesley.

Brazma, A., Jonassen, I., Vilo, J., and Ukkonen, E. (1998). Predicting gene regulatory elements in silico on a genomic scale. *Genome Research* 8:1202–1215.

Brunak, S., Engelbrecht, J., and Knudsen, S. (1990). Cleaning up gene databases. *Nature* 343:123.

Brunak, S., Engelbrecht, J., and Knudsen, S. (1990). Neural network detects errors in the assignment of mRNA splice sites. *Nucleic Acids Res.* 18:4797–4801.

Brunak, S., Engelbrecht, J., and Knudsen, S. (1991). Prediction of human mRNA donor and acceptor sites from the DNA sequence. *Journal of Molecular Biology* 220:49–65.

Brown, M. P. S., Grundy, W. N., Lin, D., Cristianini, N., Sugnet, C. W., Furey, T. S., Ares, M., and Haussler, D. (2000). Knowledge-based analysis of microarray gene expression data by using support vector machines *Proc. Natl. Acad. Sci. USA* 97:262–267.

---

[3]Accompanying web page at http://128.200.5.223/CyberT/

Burge C., and Karlin, S. (1997). Prediction of complete gene structures in human genomic DNA. *Journal of Molecular Biology* 268:78–94.

Bussemaker, H. J., Li, H., and Siggia, E. D. (2000). Building a dictionary for genomes: Identification of presumptive regulatory sites by statistical analysis. *Proc. Natl. Acad. Sci. USA* 97:10096–10100.

Chen, T., He, H. L., and Church, G. M. (1999). Modeling gene expression with differential equations *Pacific Symposium on Biocomputing* 4:29–40.[4]

Chiang, D. Y., Brown, P. O., and Eisen, M. B. (2001). Visualizing associations between genome sequences and gene expression data using genome-mean expression profiles. *Bioinformatics* 17(Suppl 1):S49–S55.

Christiansen, T., Torkington, N., and Wall, L. (1998). *Perl Cookbook*, 1st ed. Sebastopol, CA: O'Reilly & Associates.

Claverie, J.-M. (1999). Computational methods for the identification of differential and coordinated gene expression. *Hum. Mol. Genet.* 8:1821–1832.

Cole, S. T., et al. (1998). Deciphering the biology of *Mycobacterium tuberculosis* from the complete genome sequence. *Nature* 393:537–544.

D'haeseleer, P., Wen, X., Fuhrman, S., and Somogyi, R. (1999) . Linear modeling of mRNA expression levels during CNS development and injury *Pacific Symposium on Biocomputing* 4:41–52.[5]

Dudoit, S., Fridlyand, J., and Speed, T. P. (2000). Comparison of discrimination methods for the classification of tumors using gene expression data. Technical report #576, June 2000.[6]

Dudoit, S., Yang, Y., Callow, M. J., and Speed, T. P. (2000). Statistical methods for identifying differentially expressed genes in replicated cDNA microarray experiments. Technical report #578, August 2000.[7]

Dysvik, B, and Jonassen, I. (2001). J-Express: Exploring gene expression data using Java. *Bioinformatics* 17:369–70.[8]

---

[4] Available online at http://psb.stanford.edu
[5] Available online at http://psb.stanford.edu
[6] Available at http://www.stat.berkeley.edu/tech-reports/index.html
[7] Available at http://www.stat.berkeley.edu/tech-reports/index.html
[8] Software available at http://www.ii.uib.no/˜bjarted/jexpress/

Efron, B., Storey, J., and Tibshirani, R. (2001). Microarrays, empirical Bayes methods, and false discovery rates. Technical report. Statistics Department, Stanford University.[9]

Fix, E., and Hodges, J. (1951). Discriminatory analysis, nonparametric discrimination: Consistency properties. Technical report, Randolph Field, Texas: USAF School of Aviation Medicine.

Fellenberg, K., Hauser, N. C., Brors, B., Neutzner, A., Hoheisel, J. D., and Vingron, M. (2001), Correspondence analysis applied to microarray data. *Proc. Natl. Acad. Sci. USA* 98:10781–10786.

Friedman, N., Linial, M., Nachman, I., and Pe'er, D. (2000). Using Bayesian networks to analyze expression data. *Proc. Fourth Annual International Conference on Computational Molecular Biology (RECOMB) 2000.*

Fujibuchi, W., Anderson, J. S. J., and Landsman, D. (2001). PROSPECT improves cis-acting regulatory element prediction by integrating expression profile data with consensus pattern searches. *Nucleic Acids Res.* 29:3988–3996.

Gautier, L., and Knudsen, S. (2001). An R software environment for analysis of oligonucletide array data. Manuscript in preparation.

Getz, G., Levine, E., and Domany, E. (2000). Coupled two-way clustering analysis of gene microarray data. *Proc. Natl. Acad. Sci. USA* 97:12079–12084.

Goryachev, A. B., Macgregor, P. F., and Edwards, A. M. (2001). Unfolding of microarray data. *Journal of Computational Biology* 8:443–61.

Guigo, R., Knudsen, S., Drake, N., and Smith. T. (1992). Prediction of Gene Structure. *Journal of Molecular Biology* 226:141–157.

Hartemink, A. J., Gifford, D. K., Jaakkola, T. S., and Young, R. A. (2001). Using graphical models and genomic expression data to statistically validate models of genetic regulatory networks. *Pacific Symposium on Biocomputing* 6:422–433.[10]

Hacia, J. G. (1999). Resequencing and mutational analysis using oligonucleotide microarrays (Review). *Nat. Genet.* 21(1 Suppl):42–7.

Hastie, T., Tibshirani, R., Eisen, M. B., Alizadeh, A., Levy, R., Staudt, L., Chan, W. C., Botstein, D., and Brown, P. (2000). Gene shaving as a

---

[9]Manuscript available at http://www-stat.stanford.edu/~tibs/research.html

[10]Available online at http://psb.stanford.edu

method for identifying distinct sets of genes with similar expression patterns. *Genome Biol.* 1:RESEARCH0003.1–21

Hastie, T., Tibshirani, R., Botstein, D., and Brown, P. (2001). Supervised harvesting of expression trees. *Genome Biol.* 2:RESEARCH0003.

Herrero, J., Valencia, A., and Dopazo, J. (2001). A hierarchical unsupervised growing neural network for clustering gene expression patterns. *Bioinformatics* 17:126–136.

von Heydebreck, A., Huber, W., Poustka, A., and Vingron, M. (2001). Identifying splits with clear separation: A new class discovery method for gene expression data. *Bioinformatics* 17(Suppl 1):S107–S114.

Holter, N. S., Maritan, A., Cieplak, M., Fedoroff, N. V., and Banavar, J. R. (2000). Dynamic modeling of gene expression data. *Proc. Natl. Acad. Sci. USA* 98:1693–1698.

Holter, N. S., Mitra, M., Maritan, A., Cieplak, M., Banavar, J. R., and Fedoroff, N.V. (2000). Fundamental patterns underlying gene expression profiles: Simplicity from complexity. *Proc. Natl. Acad. Sci. USA* 97:8409–14.

Hu, G. K., Madore, S. J., Moldover, B., Jatkoe, T., Balaban, D., Thomas, J., and Wang, Y. (2001). Predicting splice variant from DNA chip expression data. *Genome Research* 11:1237–1245.

Hughes, T. R., Marton, M. J., Jones, A. R., Roberts, C. J., and Stoughton, R., et. al. (2000). Functional discovery via a compendium of expression profiles. *Cell* 102:109–26.

Ideker, T., Thorsson, V., Siegel, A. F., and Hood, L. (2000). Testing for differentially-expressed genes by maximum-likelihood analysis of microarray data. *Journal of Computational Biology* 7:805–817.

Ideker, T. E., Thorsson, V., and Karp, R. M. (2000). Discovery of regulatory interactions through perturbation: Inference and experimental design. *Pacific Symposium on Biocomputing* 5:305–16.[11]

Ideker, T., Thorsson, V., Ranish, J. A., Christmas, R., Buhler, J., Eng, J. K., Bumgarner, R., Goodlett, D. R., Aebersold, R., and Hood, L. (2001). Integrated genomic and proteomic analyses of a systematically perturbed metabolic network. *Science* 292:929–34.

Jarmer, H., Friis, C., Saxild, H. H., Berka, R., Brunak, S., and Knudsen, S. (2002). Inferring parsimonious regulatory networks in *B. subtilis*. *Pacific Symposium on Biocomputing* 2002. Poster presentation.

[11] Available online at http://psb.stanford.edu

Jenssen, T. K., Laegreid, A., Komorowski, J., and Hovig, E. (2001). A literature network of human genes for high-throughput analysis of gene expression. *Nat Genet.* 28:21–8.

Jensen, L. J., and Knudsen, S. (2000). Automatic discovery of regulatory patterns in promoter regions based on whole cell expression data and functional annotation. *Bioinformatics* 16:326–333.

van Kampen, A. H., van Schaik, B. D., Pauws, E., Michiels, E. M., Ruijter, J. M., Caron, H. N., Versteeg, R., Heisterkamp, S. H., Leunissen, J. A., Baas, F., and van der Mee M. (2000). USAGE: A web-based approach towards the analysis of SAGE data. *Bioinformatics.* 16:899–905.

Kerr, M. K., Martin, M., and Churchill, G. A. (2000). Analysis of variance for gene expression microarray data. *J. Comput. Biol.* 7:819–37.

Kerr, M. K., and Churchill, G. A. (2001a). Statistical design and the analysis of gene expression microarray data. *Genet Res.* 77:123–8. Review.

Kerr, M. K., and Churchill, G. A. (2001). Bootstrapping cluster analysis: Assessing the reliability of conclusions from microarray experiments. *Proc. Natl. Acad. Sci. USA* 98:8961–5.

Khan, J., Wei, J. S., Ringner, M., Saal, L. H., Ladanyi, M., Westermann, F., Berthold, F., Schwab, M., Antonescu, C. R., Peterson, C., and Meltzer, P. S. (2001). Classification and diagnostic prediction of cancers using gene expression profiling and artificial neural networks. *Nature Genetics* 7:673–679.

Kim, S., Dougherty, E. R., Bittner, M. L., Chen, Y., Sivakumar, K., Meltzer, P., and Trent, J. M. (2000) General nonlinear framework for the analysis of gene interaction via multivariate expression arrays. *J. Biomed. Opt.* 5:411–24.

Kim, S., Dougherty, E. R., Chen, Y., Sivakumar, K., Meltzer, P., Trent, J. M., and Bittner, M. (2000) Multivariate measurement of gene expression relationships. *Genomics* 15:201–9.

Knudsen, S. (1999). Promoter2.0: For the recognition of PolII promoter sequences. *Bioinformatics* 15:356–361.[12]

Knudsen, S., Nielsen, H.B., Nielsen, C., Thirstrup, K., Blom, N., Sicheritz-Ponten, T., Gautier, L., Workman, C., and Brunak, S. (2001). T cell transcriptional responses to HIV infection in vitro. Manuscript in preparation.

---

[12] Available as a web server at http://www.cbs.dtu.dk/services/Promoter/

Kohonen, T. (1995). *Self-Organizing Maps.* Berlin: Springer.

Krogh, A. (1997). Two methods for improving performance of an HMM and their application for gene finding. *Proc. Fifth Int. Conf. on Intelligent Systems for Molecular Biology (ISMB)* Menlo Park, CA: AAAI Press, pp. 179–186.

Lash, A. E., Tolstoshev, C. M., Wagner, L., Schuler, G. D., Strausberg, R. L., Riggins, G. J., and Altschul, S. F. (2000). SAGEmap: A public gene expression resource. *Genome Res.* 10:1051–60.

Lawrence, C. E., Altschul, S. F., Boguski, M. S., Liu, J. S., Neuwald, A. F., and Wootton, J. C. (1993). Detecting subtle sequence signals: A Gibbs sampling strategy for multiple alignment. *Science* 262:208–214.

Lemon, W. J., Palatini, J. T., Krahe, R., and Wright, F. A. (2001). Comparison of gene expression estimators for oligonucleotide arrays. Submitted 2001.[13]

Liu, X., Brutlag, D. L., and Liu, J. S. (2001). BioProspector: Discovering conserved DNA motifs in upstream regulatory regions of co-expressed genes. *Pacific Symposium on Biocomputing* 6:127–138.[14]

Li, C., and Wong, W. H. (2001a). Model-based analysis of oligonucleotide arrays: Expression index computation and outlier detection. *Proc. Natl. Acad. Sci. USA* 98:31–36.[15]

Li, C., and Wong, W. H. (2001b). Model-based analysis of oligonucleotide arrays: Model validation, design issues and standard error application. *Genome Biology* 2:1–11.[16]

Li, F., and Stormo, G. D. (2001). Selection of optimal DNA oligos for gene expression arrays. *Bioinformatics* 17:1067–1076.

Liang, S., Fuhrman S., and Somogyi, R. (1998). REVEAL, A general reverse engineering algorithm for inference of genetic network architectures. *Pacific Symposium on Biocomputing* 3:18–29.[17]

Lipshutz, R. J., Fodor, S. P. A., Gingeras, T. R., and Lockhart, D. J. (1999). High density synthetic oligonucleotide arrays. *Nature Genetics Chipping Forecast* 21:20–24.

[13] Preprint and Perl scripts available at http://thinker.med.ohio-state.edu/projects/fbss/index.html
[14] Available online at http://psb.stanford.edu
[15] Software available at http://www.dchip.org
[16] Software available at http://www.dchip.org
[17] Available online at http://psb.stanford.edu

Lockhart, D. J., Dong, H., Byrne, M. C., Follettie, M. T., Gallo, M. V., Chee, M. S., Mittmann, M., Wang C., Kobayashi, M., Horton, H., and Brown, E. L. (1996). Expression monitoring by hybridization to high-density oligonucleotide arrays. *Nature Biotechnology* 14:1675–1680.

Maki, Y., Tominaga, D., Okamoto, M., Watanabe, S., and Eguchi, Y. (2001). Development of a system for the inference of large scale genetic networks. *Pacific Symposium on Biocomputing* 6:446–458.[18]

Man, M. Z., Wang, X., and Wang, Y. (2000). POWER_SAGE: Comparing statistical tests for SAGE experiments. *Bioinformatics*. 16:953–9.

Margulies, E. H., and Innis, J. W. (2000). eSAGE: Managing and analysing data generated with serial analysis of gene expression (SAGE). *Bioinformatics*. 16:650–1.

Masys, D. R., Welsh, J. B., Lynn Fink, J., Gribskov, M., Klacansky, I., and Corbeil, J. (2001). Use of keyword hierarchies to interpret gene expression patterns. *Bioinformatics*. 17:319–26. [19]

Matz, M., Usman, N., Shagin, D., Bogdanova, E., and Lukyanov, S. (1997). Ordered differential display: A simple method for systematic comparison of gene expression profiles. *Nucleic Acids Res.* 25:2541–2.

Michaels, G. S., Carr, D. B., Askenazi, M., Fuhrman, S., Wen, X., and Somogyi, R. (1998). Cluster analysis and data visualization of large-scale gene expression data. *Pacific Symposium on Biocomputing* 3:42–53.[20]

Montgomery, D. C., and Runger, G. C. (1999). *Applied Statistics and Probability for Engineers*. New York: Wiley.

Neuwald, A. F., Liu, J. S., and Lawrence, C. E. (1995). Gibbs motif sampling: Detection of bacterial outer membrane protein repeats. *Protein Science* 4:1618–1632.

Newton, M. A., Kendziorski, C. M., Richmond, C. S., Blattner, F. R., and Tsui, K. W. (2001). On differential variability of expression ratios: Improving statistical inference about gene expression changes from microarray data. *J. Comput. Biol.* 8:37–52.

Nielsen, H. B., and Knudsen, S. (2001). Avoiding cross hybridization by choosing nonredundant targets on cDNA arrays. *Bioinformatics* In press.

---

[18] Available online at http://psb.stanford.edu
[19] Web-based software available at http://array.ucsd.edu/hapi/
[20] Available online at http://psb.stanford.edu

Noordewier, M. O., and Warren, P. V. (2001). Gene expression microarrays and the integration of biological knowledge. *Trends. Biotechnol.* 19:412–5.

Pan, W., Lin, J., and Le, C. (2001). How many replicates of arrays are required to detect gene expression changes in microarray experiments? A mixture model approach. Report 2001-012, Division of Biostatistics, University of Minnesota.[21]

Park, P. J., Pagano, M., and Bonetti, M. (2001). A nonparametric scoring algorithm for identifying informative genes from microarray Data. *Pacific Symposium on Biocomputing* 6:52–63. [22]

Pearson, W. R. (2001). Training for Bioinformatics and Computational Biology (Editorial). *Bioinformatics* 17:761–762.

Pe'er, D., Regev, A., Elidan, G., and Friedman, N. (2001). Inferring subnetworks from perturbed expression profiles. *Bioinformatics* 17 (Suppl. 1):S215–S224.

Piper, M., Daran-Lapujade, P., Bro, C., Regenberg, B., Knudsen, S., Nielsen, J., and Pronk, J. (2002). Reproducibility of transcriptome analyses using oligonucleotide microarrays: An interlaboratory comparison of chemostat cultures of *Saccharomyces cerevisiae*. Manuscript in preparation.

Rain, J. C., Selig, L., De Reuse, H., Battaglia, V., Reverdy, C., Simon, S., Lenzen, G., Petel, F., Wojcik, J., Schachter, V., Chemama, Y., Labigne, A., and Legrain, P. (2001). The protein-protein interaction map of *Helicobacter pylori*. *Nature* 409:211–5.

Raychaudhuri, S., Stuart, J. M., and Altman, R. B. (2000). Principal components analysis to summarize microarray experiments: Application to sporulation time series. *Pac. Symp. Biocomput.* 2000:455–66.[23]

Roberts, C. J., Nelson, B., Marton, M. J., Stoughton, R., Meyer, M. R., Bennett, H. A., He, Y. D. D. , Dai, H. Y., Walker, W. L., Hughes, T. R., Tyers, M., Boone, C., and Friend, S. H. (2000). Signaling and circuitry of multiple MAPK pathways revealed by a matrix of global gene expression profiles. *Science* 287:873–880.

Samsonova, M. G., and Serov, V. N. (1999). NetWork: An interactive interface to the tools for analysis of genetic network structure and dynamics. *Pacific Symposium on Biocomputing* 4:102–111.[24]

---

[21] Available at http://www.biostat.umn.edu/cgi-bin/rrs?print+2001
[22] Manuscript available online at http://psb.stanford.edu
[23] Available online at http://psb.stanford.edu
[24] Available online at http://psb.stanford.edu

Sasik, R., Hwa, T., Iranfar, N., and Loomis, W. F. (2001). Percolation clustering: A novel algorithm applied to the clustering of gene expression patterns in dictyostelium development. *Pacific Symposium on Biocomputing* 6:335–347.[25]

Segal, E., Taskar, B., Gasch, A., Friedman, N., and Koller, D. (2001). Rich probabilistic models for gene expression. *Bioinformatics* 17(Suppl 1):S243–S252.

Schadt, E. E., Li, C., Su, C., and Wong, W. H. (2000). Analyzing high-density oligonucleotide gene expression array data. *J. Cell. BioChem.* 80:192–201.

Schena, Mark. (1999). *DNA microarrays: A practical approach* (Practical Approach Series, 205). Oxford: Oxford Univ. Press.

Schena, Mark. (2000). *Microarray Biochip Technology*. Sunnyvale, CA: Eaton.

Scherf, M., Klingenhoff, A., and Werner, T. (2000). Highly specific localization of promoter regions in large genomic sequences by PromoterInspector: A novel context analysis approach. *Journal of Molecular Biology* 297:599–606.

Schuchhardt, J., Beule, D., Malik, A., Wolski, E., Eickhoff, H., Lehrach, H., and Herzel, H. (2000). Normalization strategies for cDNA microarrays. *Nucleic Acids Res.* 28:E47.

Schwartz, R. L. and Phoenix, T. (2001). *Learning Perl*, 3rd ed. Sebastopol, CA: O'Reilly & Associates

Skovgaard, M., Jensen, L. J., Brunak, S., Ussery, D., and Krogh, A. (2001). On the total number of genes and their length distribution in complete microbial genomes. *Trends Genet.* 17:425–428.

van Someren, E. P., Wessels, L. F. A., and Reinders, M. J. T. (2000). Linear modeling of genetic networks from experimental data. *Proc. ISMB* 2000:355–366.

Spellman, P., Sherlock, G., Zhang, M., Lyer, V., Anders, K., Eisen, M., Brown, P., Botstein, D., and Futcher, B. (1998). Comprehensive identification of cell cycle-regulated genes of yeast *S. cerevisiae* by microarray hybridization. *Mol. Biol. Cell* 9:3273–3297.

Spicker, J. S., Wikman, F, Lu M. L., Cordon-Cardo C., Ørntoft, T. F., Brunak, S., and Knudsen, S. (2001). Neural network trained to predict TP53 gene sequence based on p53. Manuscript in preparation.

---

[25] Available online at http://psb.stanford.edu

Szallasi, Z. (1999). Genetic network analysis in light of massively parallel biological data acquisition. *Pacific Symposium on Biocomputing* 4:5–16.[26]

Tamayo, P., Slonim, D., Mesirov, J., Zhu, Q., Kitareewan, S., Dmitrovsky, E., Lander, E. S., and Golub, T. R. (1999). Interpreting patterns of gene expression with self-organizing maps: Methods and application to hematopoietic differentiation. *Proc. Natl. Acad. Sci. USA* 96:2907–2912.

Tanabe, L., Scherf, U., Smith, L. H., Lee, J. K., Hunter, L., and Weinstein, J. N. (1999). MedMiner: An internet text-mining tool for biomedical information, with application to gene expression profiling. *BioTechniques* 27:1210–1217.[27]

Tanay, A., and Shamir, R. (2001). Expansion on existing biological knowledge of the network: Computational expansion of genetic networks. *Bioinformatics* 17(Suppl 1):S270–S278.

Thieffry, D., and Thomas, R. (1998). Qualitative analysis of gene networks. *Pacific Symposium on Biocomputing* 3:77–88.[28]

Tibshirani, R., Walther, G., Botstein, D., and Brown, P. (2000). Cluster validation by prediction strength. Technical report. Statistics Department, Stanford University.[29]

Thomas, J. G., Olson, J. M., Tapscott, S. J., and Zhao, L. P. (2001). An efficient and robust statistical modeling approach to discover differentially expressed genes using genomic expression profiles. *Genome Res.* 11:1227–1236.

Thykjaer, T., Workman, C., Kruhøffer, M., Demtröder, K., Wolf, H., Andersen, L. D., Frederiksen, C. M., Knudsen, S., and Ørntoft, T. F. (2001). Identification of gene expression patterns in superficial and invasive human bladder cancer. *Cancer Research* 61:2492–2499.

Tusher, V. G., Tibshirani, R., and Chu, G. (2001). Significance analysis of microarrays applied to the ionizing radiation response. *Proc. Natl. Acad. Sci. USA* 98:5119–5121.[30]

Varotto, C., Richly, E., Salamini, F., and Leister, D. (2001). GST-PRIME: A genome-wide primer design software for the generation of gene sequence tags. *Nucleic Acids Res.* 29:4373–7.

---

[26] Available online at http://psb.stanford.edu
[27] Web version available at http://discover.nci.nih.gov/textmining/filters.html
[28] Available online at http://psb.stanford.edu
[29] Manuscript available at http://www-stat.stanford.edu/~tibs/research.html
[30] Software available for download at http://www-stat.stanford.edu/~tibs/SAM/index.html

Velculescu, V. E., Zhang, L., Vogelstein, B., and Kinzler, K. W. (1995). Serial analysis of gene expression. *Science.* 270:484–7.

Venables, W. N. and Ripley, B. D. (1999). *Modern applied statistics with S-PLUS*, 3rd ed. New York, NY: Springer.

Vingron, M. (2001). Bioinformatics Needs to adopt statistical thinking (Editorial). *Bioinformatics* 17:389–390.

Wahde, M., and Hertz, J. (2000). Coarse-grained reverse engineering of genetic regulatory networks. *Biosystems* 55:129–36.

Wahde, M., and Hertz, J. (2001). Modeling genetic regulatory dynamics in neural development. *J. Comput. Biol.* 8:429–42.

Wall, L., Christiansen, T., and Orwant, J. (2000). Programming Perl, 3rd ed. Sebastopol, CA: O'Reilly & Associates.

Wall, M. E., Dyck, P. A., and Brettin, T. S. (2001). SVDMAN—singular value decomposition analysis of microarray data. *Bioinformatics* 17:566–568.

Weaver, D., Workman, C., and Stormo, G. (1999). Modeling regulatory networks with weight matrices. *Pacific Symposium on Biocomputing* 4:122–123.[31]

Wessels, L. F. A., Van Someren, E. P., and Reinders, M. J. T. (2001) A Comparison of genetic network models. *Pacific Symposium on Biocomputing* 6:508–519.[32]

Wikman, F. P., Lu, M. L., Thykjaer, T., Olesen, S. H., Andersen, L. D., Cordon-Cardo, C., and Ørntoft, T. F. (2000). Evaluation of the performance of a p53 sequencing microarray chip using 140 previously sequenced bladder tumor samples. *Clin. Chem.* 46:1555–61.

Wodicka, L., Dong, H., Mittmann, M., Ho, M. H., and Lockhart, D. J. (1997). Genome-wide expression monitoring in Saccharomyces cerevisiae. *Nature Biotechnology* 15:1359–1367.

Wolfsberg, T. G., Gabrielian, A. E., Campbell, M. J., Cho, R. J., Spouge, J. L., and Landsman, D. (1999). Candidate regulatory sequence elements for cell cycle-dependent transcription in *Saccharomyces cerevisiae. Genome Res.* 9:775–792.

---

[31] Available online at http://psb.stanford.edu
[32] Available online at http://psb.stanford.edu

Workman, C., and Stormo, G.D. (2000) ANN-Spec: A method for discovering transcription factor binding sites with improved specificity. *Pacific Symposium on Biocomputing 2000.*[33]

Workman, C., Jensen, L. J., Jarmer, H., Saxild, H. H., Berka, R., Gautier, L., Nielsen, H. B., Nielsen, C., Brunak, S., and Knudsen, S. (2001). A new non-linear method to reduce variability in DNA microarray experiments. Manuscript submitted.[34]

Xia, X, and Xie, Z. (2001). AMADA: Analysis of microarray data. *Bioinformatics* 17:569–70.

Xing, E. P., and Karp, R. M. (2001). CLIFF: Clustering of high-dimensional microarray data via iterative feature filtering using normalized cuts. *Bioinformatics* 17(Suppl 1):S306–S315.

Xiong, M., Jin, L., Li, W., and Boerwinkle, E. (2000). Computational methods for gene expression-based tumor classification. *Biotechniques* 29:1264–1268.

Yamamoto, M., Wakatsuki, T., Hada, A., and Ryo, A. (2001). Use of serial analysis of gene expression (SAGE) technology. *J Immunol. Methods.* 250:45–66. Review.

Yeang, C. H., Ramaswamy, S., Tamayo, P., Mukherjee, S., Rifkin, R. M., Angelo, M., Reich, M., Lander, E., Mesirov, J., and Golub, T. (2001). Molecular classification of multiple tumor types. *Bioinformatics* 17(Suppl 1):S316–S322.

Yeung, K. Y., Fraley, C., Murua, A., Raftery, A. E., and Ruzzo, W. L. (2001) Model-based clustering and data transformations for gene expression data. *Bioinformatics* 17:977–87.

Yeung, K. Y., Haynor, D. R., and Ruzzo, W. L. (2001). Validating clustering for gene expression data. *Bioinformatics* 17:309–18.

Zien, A., Aigner, T., Zimmer, R., and Lengauer, T. (2001). Centralization: A new method for the normalization of gene expression data. *Bioinformatics* 17(Suppl 1):S323–S331.

Zhao, L. P., Prentice, R., and Breeden, L. (2001). Statistical modeling of large microarray data sets to identify stimulus-response profiles. *Proc. Natl. Acad. Sci. USA* 98:5631–5636.

---

[33] Available online at http://psb.stanford.edu
[34] Software available at http://www.cbs.dtu.dk/biotools/oligoarray/

Zhu, J., and Zhang, M. Q. (2000). Cluster, function and promoter: Analysis of yeast expression array. *Pacific Symposium on Biocomputing* 5:476–487.[35]

---

[35] Available online at http://psb.stanford.edu

Index